T0073903

Scientific History

Scientific History

Experiments in History and Politics from the Bolshevik Revolution to the End of the Cold War

Elena Aronova

The University of Chicago Press :: Chicago and London

The University of Chicago Press, Chicago 60637
The University of Chicago Press, Ltd., London

Published 2021
Printed in the United States of America

30 29 28 27 26 25 24 23 22 21   1 2 3 4 5

ISBN-13: 978-0-226-76138-1 (cloth)
ISBN-13: 978-0-226-76141-1 (e-book)
DOI: https://doi.org/10.7208/chicago/9780226761411.001.0001

Library of Congress Cataloging-in-Publication Data

Names: Aronova, E. A. (Elena Aleksandrovna), author.
Title: Scientific history : experiments in history and politics from the Bolshevik revolution to the end of the Cold War / Elena Aronova.
Description: Chicago ; London : The University of Chicago Press, 2021. | Includes bibliographical references and index.
Identifiers: LCCN 2020038272 | ISBN 9780226761381 (cloth) | ISBN 9780226761411 (ebook)
Subjects: LCSH: Historiography—History—20th century. | Historiography—Europe—History—20th century. | History—Methodology—History. | Science—History—Historiography.
Classification: LCC D13 .A76 2021 | DDC 907.2—dc23
LC record available at https://lccn.loc.gov/2020038272

♾ This paper meets the requirements of ANSI/NISO Z39.48-1992 (Permanence of Paper).

Contents

Preface

This book is a history, but it was shaped by my experiences as a scientist. My undergraduate training and first career were in science. I graduated from the Chemistry Department of Moscow State University in 1991, precisely at the moment that the country that had provided my education and guaranteed my employment ceased to exist. I was fortunate to land a position at the Institute of Bioorganic Chemistry in Moscow, a leading institute in my field. As I began work as a biochemist in the laboratory of Gene Bioengineering, the economy was melting all around me. Libraries discontinued costly journal subscriptions, labs stopped receiving reagents, and many institutes in Moscow could not even afford electricity. Any time colleagues traveled abroad to attend a conference or do research, they returned with reprints and lab supplies. Most often, they would bring a bagful of Eppendorf tubes—an indispensable laboratory consumable for working with small sample preparations, from 0.5 to 2 milliliters. Eppendorf tubes are intended to be used one time only, but, out of necessity, we washed and reused them many times. It was not a good time to do molecular biology in Russia.

After a couple of years, I began considering other options. I took a friend's offer to kickstart a lab that applied the then-new technique of the polymerase chain reaction, or PCR, to detect sexually transmitted diseases in screening tests administered through the Russian Ministry of Public Health. It was the first lab of this kind in Moscow, and the job paid well, but my colleagues and I still lacked basic gear and washed our Eppendorfs, which risked contaminating the test samples. Given the stakes involved, a do-it-yourself approach to molecular diagnostics felt scary and very wrong.

It was around this time that I first encountered the history of science. Initially, I thought of the history of science as a way to continue doing science, just by other means—it seemed to be an ideal compromise solution in my situation. I entered a doctoral program at the Institute for the History of Science and Technology in Moscow while continuing to work in a wet lab and wrote a dissertation on the conceptual history of immunology, earning a degree in biology. By then, however, I had become genuinely interested in history and ended up pursuing another doctorate, this time in history, at the University of California, San Diego.

My journey from the lab to the archives made me observant of parallels between scientific method and historical method. As a scientist, I had collected, processed, described, compared, and interpreted data. I backed my analysis with evidential support to corroborate my conclusions. Historians do pretty much the same. I was curious to find out whether the parallels extend further than that. This book is my attempt to think through a history of disciplinary crossovers by mapping out various ways in which historians and scientists have exchanged and shared methodologies, approaches, and subject matters.

Today, a growing number of scholars in the humanities are calling for a reengagement with the natural sciences. Taking their cues from recent breakthroughs in genetics and the neurosciences, the advocates of "biohistory" and "deep history" call for a reassessment of long-held assumptions about the very definition of history, its methods, and its evidentiary base.[1] The most polemical of the present-day revisionist projects, Big History, seeks to unite what C. P. Snow called the "two cultures" of the sciences and the humanities under the umbrella of the history of the universe, from the big bang to the present day, putting human history in the context of cosmic history and historians into conversation with practitioners in biology, geology, and other sciences.[2] Digital humanities is another area in which the two cultures of the sciences and the humanities have come together after a period of what was perceived as a separation. After the cultural, linguistic, transnational, and other historiographic "turns" that marked the writing of history in the later part of the twentieth century, the historiography is, as some claim, experiencing a "scientific turn" in the first decades of the twenty-first century.

One of the things I learned in my professional training as a historian is that historiographic turns often have longer histories than their claims of novelty would suggest and that historicizing the turns brings to the fore a more sophisticated inventory of approaches, possibilities, and intellectual trends than turning talk alone allows.[3] Looking back at their own past, contemporary historians recognize that their nineteenth-century counterparts shared interests and methodologies with natural scientists. These connections have, however, become submerged in historians' accounts of their own discipline in the twentieth century. This collective amnesia makes the present-day scientific turn in history appear revolutionary and new. But, as the account I provide in this book shows, there was a scientific turn in history at every turn since history was established as a discipline.

There was, of course, a scientific turn in the late nineteenth century when Leopold von Ranke's seminars in Berlin established scientific standards of evidence as a norm of professional historical research. But there was also a scientific turn driven by biology, first in the form of nineteenth-century movements to analogize biological evolution and historical development, and then in the form of early twentieth-century attempts to use genetics as a resource to overcome the historical determinism inherent in evolutionary explanations of history. In the first decades of the twentieth century, many scientists and historians regarded the history of science as a bridge between the humanities and the sciences, and it was practiced accordingly. In the 1940s and 1950s, the scientific turn was driven by a demand for a less Eurocentric and more global world history. In the following decades, the arrival of computers prompted new practices of data mining and promoted already-existing quantitative conventions and practices of collecting, aggregating, and processing data. All these can legitimately be viewed as scientific turns. None, however, figure in contemporary historians' calls to rearticulate their inquiries as enterprises "distinct" from "yet complementary to" those in the natural sciences.[4] This book maps out the submerged history of historians' continuous engagement with the methods, tools, and values of the natural sciences throughout the twentieth century. I call this engagement *scientific history*.

The history of scientific history, in all its forms, would be a very big story. The book is necessarily selective and does not claim to offer a comprehensive account of scientific history in all its different variations. Rather, this is an outline history of how people have thought about what perspectives from genetics, botany, or computer sciences could do for history as well as about what history could do for these sciences and why scholars should be concerned with these matters. In the same way that my motivations for this project were driven by my own trajectory and experiences, I ground this historical account in the life trajectories of individuals. Many of them, such as the historians associated with the *Annales* school of historiography and the biologists Julian Huxley, Nikolai Vavilov, and John Desmond Bernal, will be familiar to readers from existing historiographic literature. However, the interactions *between* these historians and these and other biologists have escaped the attention of the historians and receded from our historical memory. I argue that these interactions drove the scientific turns of their time.

My points of departure and arrival for this story are also a reflection of my own personal and professional trajectory, which took me from Russia to the United States via extended stays in France and Germany. I was naturally interested in these countries' cultural outlooks and the experiences of individuals involved in similar cross-cultural exchanges. All the stories in this book involve transnational encounters, particularly French scholars' interactions with Soviet scientists. These encounters, I found, profoundly shaped the writing of history in the twentieth century, but the Soviet part of the story has become separated from the history of historiographic thought by an intellectual iron curtain. More often than not, however, it is the submerged parts of history that are most interesting. As we shall see, the story of scientific history looks very different if we consider the histories of historiographic thought and the histories of science in Russia/the Soviet Union together.

Introduction

The end of 1931 found Arnold Toynbee completing that year's volume of the *Survey of International Affairs*, a monumental year-by-year study of world politics. He described the year that had just passed as an "*annus horribilis*," a year in which "men and women all over the world were seriously contemplating and frankly discussing the possibility that the Western system of society might break down and cease to work."[1] Simultaneously, the historian was working on another monumental project, *A Study of History*. Combining a global view on history with recourses to archaeology, sociology, biology, anthropology, linguistics, paleontology, and other sciences, the *Study* was to provide a window into different "fossil civilizations"—the Minoan, Sumerian, Hittite, Babylonic, Andean, Mayan, Yucatec, and Mexic cultures—that had gone through different "times of trouble." It was his hope that a study of these civilizations—their genesis, growth, breakdown, and disintegration—could offer insights into the causes and consequences of the "time of trouble" of his own time.[2]

For much of the twentieth century, Toynbee was one of the world's most read, translated, and discussed historians. In equal measure to his success with the general readership, he drew fire from his fellow historians, who mocked his moralizing style and denounced his method.[3]

Yet, many of his critics shared his conviction that understanding both the past and the present required a perspective that would draw on the resources of all disciplines. Not unlike Toynbee, many thought that achieving social stability hinged on social unity, which, in turn, hinged on achieving intellectual unity across the sciences and the humanities.

This book examines several waves of different kinds of experimentation with the scale and method of history in the twentieth century, each of which surged highest at perceived times of trouble, from the crises-ridden decades leading to the First World War to the Cold War. I ground this history in the intertwined trajectories of six intellectuals and the larger programs they set in motion: Henri Berr (1863–1954), a teacher of rhetoric at the elite Lycée Henri IV in Paris; Nikolai Bukharin (1888–1938), a Bolshevik politician whose tragic finale inspired Arthur Koestler's famous political thriller *Darkness at Noon* (1940); Lucien Febvre (1878–1956), a French historian best known as a cofounder of the *Annales* school of historiography; Nikolai Vavilov (1887–1943), a Soviet geneticist who shared Bukharin's fate few years later; and two maverick British biologists, Julian Sorell Huxley (1887–1975) and John Desmond Bernal (1901–71), who lived long enough to experience two world wars and, at the height of the Cold War, the acute possibility of a third.

What could these individuals have in common, except for the fact that they all were men and held powerful positions during their lifetime? These men, indeed, came from different backgrounds, pursued different goals, held different political views, and expressed themselves in different mother tongues. Yet they can serve as representatives of a larger international motley crew of scientists, historians, journalists, activists, entrepreneurs, and public figures who sought to reexamine the boundaries, tools, and uses of history and who created powerful institutions and networks to support their projects. The fact that they all were men suggests that these grand programs have a gendered dimension as well.[4]

These six men also met each other in person, or at the very least they were present in the same places at the same times and for the same reasons, as participants in the international congresses in the nascent field of the history of science. Berr, Bukharin, Vavilov, Huxley, and Bernal all attended the Second International Congress of the History of Science and Technology, held in London in that annus horribilis, 1931. Berr himself hosted the first congress, held two years earlier at the International Center of Synthesis (Centre international de synthèse) in Paris, the institutional home of his program of "historical synthesis." Febvre, who codirected the center with Berr, was involved with the organization and the proceedings of that congress. Amid the unprecedented social and political instability that unfolded throughout Europe and the United States between 1929 and 1931, these scholars regarded the history of science as a special scholarly enterprise with double loyalties: to science and to history. In Berr's words, the goal of the history of science was "to establish a tight liaison between the sciences of nature and those of the humanities."[5]

The different protagonists of this book envisioned and enacted the "liaison" between history and science differently. In the first two chapters, I follow Berr (chapter 1) and Bukharin (chapter 2) and their programs of historical synthesis. In the decades around 1900, each reconciled history and science within competing frameworks of the unity of knowledge rooted in two major intellectual formations of the nineteenth century, positivism and Marxism, as manifested in their application to history and its method. The new venue of international congresses of historians became a stage on which a high drama of performative confrontation between these programs (i.e., a principally French refurbishing of Comte's positive philosophy and a principally Russian/early Soviet refurbishing of Marxist philosophy) played out in the early 1930s, as I discuss in chapter 2. The encounters at one of these events, the 1931 London congress, set the stage for the rest of the book.

In chapter 3, I follow Bukharin's protégé Vavilov and his work on the centers of origin of cultivated plants—work he presented at the 1931 London congress. In the aftermath of the congress, Berr's protégé Febvre publicized Vavilov's work in the pages of the *Annales*, as an approach that reconciled biology and history through genetics. This chapter examines how the specific content of science—Vavilov's genogeography—became a resource for historians associated with the *Annales* school, who were building, at the same time, on the longer tradition of bridging biology and history.

Chapters 4–6 trace some of the threads that started at the London congress into the post–World War II era. They demonstrate that, while the history of science's institutional and intellectual independence from the field's earlier roots in science became a defining feature of its identity, the romance between history and science continued at the margins of the field. In the late 1940s, at the onset of the Cold War, Febvre joined forces with another congress participant, the biologist Julian Huxley, when he became the head of the newly established UNESCO. Together, they launched what eventually became the "History of Mankind: Cultural and Scientific Development" project, which was aimed at producing a comprehensive history of the modern world in which the history of science was assigned a primary role. Huxley regarded it as a "key project" of UNESCO's. Chapter 4 discusses how he ended up in this unlikely position by tracing his trajectory between the London congress and the launch of the History of Mankind project.

The implementation of the UNESCO history project is examined in detail in chapter 5. One of the distinct features of the project was its organization as a collaborative research endeavor involving hundreds of participants worldwide, emulating a mode of research common in science but extremely rare in the humanities. While the collaborative organization was touted by its leaders as an unprecedentedly novel project, both Huxley and Febvre drew on their previous experiences of team research to design the minutiae of the project's implementation. Moreover, a Soviet team, which joined the project in 1955, brought with it the experience of nationwide coordinated research in history. The international, multilingual journals created to manage and organize the international collaboration of historians associated with the project became a medium through which work that later became central for successive generations of world historians circulated for the first time.

In this later period, as I discuss in chapter 6, the history of science became one of the areas of history in which scholars began using electronic computers and computer-enabled data analytics in their historical work. As in the other chapters, I trace these developments back to the 1931 London congress. I examine the linkages between computers, datafication, and the writing of history by following another congress participant, J. D. Bernal, and his lifelong campaign for a revolutionary, anticapitalist science shaped by both informationist and socialist visions. Bernal's program interestingly intersected with the trajectory of the Philadelphia entrepreneur Eugene Garfield, who pitched the Science Citation Index, a data-analytic tool, to the historians of science while envisioning history as a data science.

Scientific history is my shorthand for the diverse ways in which scientists and historians reconciled the techniques, approaches, and values of science with the writing of history, particularly the history of science. Whether by using tools and evidence from the contemporary sciences as their resources or by engaging in collaborative research and writing, the scholars discussed in this book not only crossed disciplinary boundaries between the sciences and the humanities but also consistently reflected on their practice of making scientific history.

As a term, *scientific history* most commonly refers to the historiographic school of Leopold von Ranke, which was instrumental in establishing standards of the history profession on the basis of precise research and careful criticism of documents found in archives, the practice that became a signifier of professional history beginning in the late nineteenth century.[6] The term is used in a wider sense as well. The adjective *scientific* is a common synonym for *objective*, or possibly *scientistic*, when referring to faith in an ideal of objectivity made possible through science or describing a futile hope of uncovering laws to explain society and its history just as astronomers explained the motions of the planets.[7] To add to the ambiguities, the adjective *scientific* also has different connotations in different languages. In Germanophone, Francophone, and Slavonic academic traditions, history counts as one of the sciences, and the distinction between the humanities and the sciences is less hardened than it is in the Anglophone academy.

Despite the multiple meanings of the term, there is a good rationale for sticking to *scientific history* as a label for the programs and activities discussed in this book. When the history of science emerged as a recognizable field in the first decades of the twentieth century, its practitioners used the terms *history of science* and *scientific history* interchangeably.[8] As an actors' category, it underscored the early vision of the field as hybrid and cross-disciplinary, implying a historical investigation of science's past, by either science practitioners or historians, as well as a larger vision of the field as a liaison between the sciences and the humanities, a vision that was enacted via a wide range of attitudes and practices.[9] Even though the meaning and the disciplinary visions of the history of science have narrowed with time, we can trace a coherent lineage of historians and scientists associated with history of science communities who sought to integrate the ideas, techniques, practices, and values of different contemporary sciences as resources for historical understanding. This book recovers this submerged school of thought.[10]

Today, when the all-encompassing scale of the anthropogenic environmental change transforming every aspect of life on earth has become undeniable, many historians are compelled to articulate the implications of the Anthropocene for the practice of history in the twenty-first century.[11] As the historian Dipesh Chakrabarty has pointed out, by placing our present not in the context of recent human history but in the context of geologic time, the concept of the Anthropocene challenges the mode of thinking about historical time that historians are used to and often take for granted.[12] In an influential essay, he has argued that "anthropogenic explanations of climate change spell the collapse of the age-old humanist distinction between natural history and human history."[13]

Whether or not historians and other scholars in the humanities agree on the use of the popular, though controversial, term *Anthropocene*, the concept it stands for—the recognition that unprecedented global environmental change is transforming the earth and human society alike—is shaping the practice of history in the twenty-first century in crucial ways. Increasingly, historians are questioning professional conventions about the legitimate scale of history and the commensurability, or the lack thereof, between the perspectives, approaches, and methodologies of history and those of the natural sciences that have been in place since history became a professional academic discipline in the late nineteenth century.[14] Programs within the historical profession, such as Big History, deep history, and biohistory as well as other emerging science-oriented approaches within the history profession, are trying to articulate modes of constructive engagement with the natural sciences.[15] But these programs, too, have a past. It is time to begin situating the history of bigger and deeper approaches to history within the longer trajectory of similar efforts across the sciences and the humanities. Historians of science such as Marianne Sommer, Nasser Zakariya, and Deborah Coen, among others, have begun to chart this vast territory, offering critical insights onto today's scientific turn in history and the humanities in general.[16]

Building on these insights, I argue that the history of the history of science *itself* is instructive for today's repositioning of history vis-à-vis the sciences, in that it reveals different interdisciplinary contexts in which historians and natural scientists interacted, the techniques that enacted such interaction, and the political uses to which these programs were put. As the field became solidly rooted in the history departments of the North American academy, the history of science triumphantly distanced itself from its early roots in the sciences.[17] Yet one of the takeaways from this book is that the discipline's current embrace of historicity does not conflict with its founders' vision of a great humanistic program committed to reconciling historical understanding and scientific explanation.[18] The submerged histories traced in this book—which unfolded in the liminal spaces on the margins of disciplines and across or outside conventional dichotomies of power—reveal continuous and complicated relationships between history and science that were diverse, ambiguous, and, at times, surprisingly productive.[19]

Russia as Method

Historians have long argued that a region can be used strategically as a methodological opportunity.[20] In science and technology studies, scholars have been discussing the concept of "Asia as method"—a shorthand for an approach that uses *Asia* as an anchoring point from which to multiply frames of reference in discussions of global modernity and intervene methodologically in the history of science, a field that originally had its primary mooring in European intellectual traditions.[21] In this approach, "the multiplicity, ambiguity, and elusiveness of 'Asia'" are being strategically deployed as useful heuristics by scholars seeking to "appropriate richer historical resources and more coherent ideological apparatuses than some other critical frameworks, such as the unwieldy 'global south.'"[22] Scholars studying regions beyond the Asian continent such as Latin America and Africa as well as the Atlantic world and the Indian Ocean world—the geographic and geopolitical spaces beyond traditional regions—have anticipated this line of reasoning and have similarly used a geographic lens to deuniversalize, regionalize, or otherwise subvert the Eurocentric history of the world.[23]

By analogy with the research strategy of Asia as method, I use Russia strategically as a method, or a heuristic, to deuniversalize history's own analytic lenses.[24] I consider several trends in historiography, such as the *Annales* school, quantitative history, and world history, using Russia as an anchoring point that reveals the circulation, appropriation, and modification of knowledge, practices, and philosophies associated with scientific history.[25] From the latter part of the nineteenth century and throughout most of the twentieth century, Russia, and then the Soviet Union, was an active, although unequal, participant in the quest for scientific history, and this in turn contributed to the reconfiguration of knowledge in the sites of its origin.

I use Asia as method as my point of departure, partly because of its explanatory power and partly because there are clear historical parallels and connections between Asia and Russia, both in terms of their geographies and in terms of their histories. The historical imagination of both regions is shot through with the notion of a universal, essentialized "West" that served and continues to serve as a cultural compass for the intellectuals and reformers in Asia and Russia alike.[26] Throughout Russian history, the West had been Russia's most important existential Other—a point of reference, a goal to catch up with and emulate, an object of both desire and resentment. But, while the various societies and cultures of the vast territory of Asia developed their own cultural, intellectual, and economic spaces, Russia has been part of the European cultural tradition, in which its elites were fully integrated. Its imperial experience was distinctly ambiguous and hybrid, occupying a liminal space between colonizer and colonized, West and non-West, Orientalist and Oriental. As Alexander Etkind has argued, Russia's experience of "internal colonization" made Russian culture "in its different aspects and periods . . . both the subject and object of orientalism."[27] In this view, Russia has colonized itself and has been Orientalized by its own outsiders—the Europeanized upper classes and elites.

The perception of Russia as a colony of Europe, colonized from the inside, was the key point of contention between the so-called Slavophiles and the Westernizers in imperial Russia.[28] It was brought up again in the early years of Soviet rule by the Marxist scholar Mikhail Pokrovskii and later by Joseph Stalin (I discuss the historical context in which these arguments were brought up and their significance in Soviet cultural politics in chapter 2).[29] Ambiguities abound in the Soviet era. The Soviet "great world-historic antithesis," to quote Nikolai Bukharin's presentation at the 1931 London congress, did not challenge the Western world order but, rather, claimed to represent a legitimate path for how the world should evolve.[30] In this geopolitical vision, Asia played a central role, as I show in chapters 3 and 5.

Russia's ambiguous position between its Occidentalism and its own Orientalized periphery and between world capitalism and the Soviet project of building world socialism defies binary oppositions between the West and its Others by positioning them as analytically connected. For these reasons, Russia is good to think with and think from about questions usually approached from the perspective of West-centered intellectual history. The history of history itself, as it developed from a literary vocation to a leading discipline in colleges and universities worldwide, is a case in point. Russia as method helps reveal the epistemological limitations of imposing the West as a universal method to the study of historiographic thought and historical methodologies that developed through exchanges and mutual appropriations across time and space.

Since the beginning of the 1990s, intellectual historians, like many other historians, began applying international, transnational, and global lenses to their subject matter.[31] Few histories of the Russian, let alone the Soviet, *gumanitarnye nauki* (which stands for both the humanities and the human sciences in Russian), however, offered transnational stories.[32] As Oleg Kharkhordin has succinctly articulated the issue: "Russians did not offer the world their own Wittgensteins, Malinowskis, Hayeks, Polanyis, and Freuds." In the Soviet Union, the humanities and the social sciences were equated with Marxism-Leninism and subjected to ideological censorship and oversight. As Kharkhordin puts it: "When the closed universe of Soviet social sciences opened to the world, it proved to be largely empty by that world's standards." Consequently, he argues, Russian scholarship in the humanities came to function as a resource-based economy: methods and theory have been "import[ed] . . . into

Russia" from Western Europe and North America, with mainly the "raw material" (such as data mined in Russian archives) being exported back. While there was an outflow of talent from Russia to the rest of the world, Kharkhordin points out, the flow of *knowledge* was for the most part unidirectional, going from the West to the East.[33] In historiography, for instance, beginning in the 1970s, an increasing number of translations brought Anglophone, Francophone, and Germanophone historiographic methods and approaches to the Soviet Union and, after its dissolution, to post-Soviet Russia. A prime example is the *Annales* school of historiography, which was popularized in the Soviet Union by the medievalist Aaron Gurevich through translations in the 1970s and became one of the leading trends in post-Soviet historiography in the 1990s.[34] With the exception of a few stars such as Mikhail Bakhtin and Yuri Lotman, however, Soviet scholarship in the humanities has not been "exported back" and integrated "into the world's knowledge-producing markets."[35]

In this book, I do not see the process of interactions between scholars in Russia and their counterparts in Western Europe and North America as either linear or one that could be described in terms of the binary import/export. Rather, I see it as complex, innovative, and creative, spanning the multiple worlds and discourses that shaped and influenced the West itself. The case of the *Annales* historians' interactions with Vavilov in the 1930s, discussed in chapter 3, is but one example illustrating the point. I show that historians associated with the *Annales* school have followed his genogeographic research, using it as a resource in their own work. They not only popularized his work in France but also established direct contact with his Institute of Plant Industry in Leningrad. The connections between French historians and Soviet scholars and scientists resumed in the 1950s in the context of the UNESCO scientific and cultural history project, as I discuss in chapter 5. Scientists and scholars in the Soviet Union had, in other words, been collaborating with the *Annales* group since its inception. These exchanges across disciplinary and political borders had been taking place for decades before Gurevich popularized the *Annales* approach in the Soviet Union in the 1970s. My point is not to suggest that Vavilov's or other Russians' ideas somehow influenced the *Annales* school but rather to undermine an understanding of knowledge as traveling in a unilinear direction from the centers of production to the rest of the world.[36]

By pushing back against the notion of a linear and unidirectional spread of knowledge, and by shifting the focus from disciplinary knowledge to the unifying problem of "scientific humanities," the book recovers the intertwined histories of the sciences and the humanities within the Russian/Soviet scientific landscape. Only by unearthing such histories can we reconnect the disciplinary histories of the sciences and those of the humanities.

A larger argument of this book is that, while these various programs of scientific history were driven by genuinely intellectual questions, they all also had powerful political drivers. In the aftermath of the Bolshevik Revolution in Russia, the champions of the revolutionary, socialist science lauded the interdisciplinary, revolutionary science under socialism as a counterideal to a capitalistic science. They contrasted it with the specialization and fragmentation of academic disciplines that were, they argued, the defining feature of individualistic, "bourgeois science." Accessible language, another defining feature of socialist science, was promoted as a means of communication across disciplinary boundaries. In this context, Ivan Borichevsky, the philologist working at the Communist Academy in the 1920s, proposed a new "science of science" that he called *naukovedenie* (lit. "science studies"). Lamenting "a growing division of labor in the sciences," with "science more and more resembling a huge industrial factory where each worker performs a particular task," Borichevsky presented a case for *naukovedenie* as a metalanguage of communication between different "scientific workshops."[37] It was in this context as well that Bukharin extolled the history of science as a tool of disciplinary border crossing and communication, commissioning what became Vavilov's most elaborate statement on the historical implications of his genogeographic research.

During the Cold War, politics and ideology shaped the programs of scientific history on both sides of the political divide. For Huxley, for example, it was Vavilov's tragic downfall as a victim of Lysenko's attacks on genetics and, more broadly, what he regarded as dangerous antiscience ideology taking roots in the Soviet Union that turned him from an early enthusiastic observer of the Soviet scientific socialist "experiment" in action to an ideological Cold Warrior who repackaged scientific history as a Western counterpart to Soviet universalist ideology. The meaning of the Soviet experiment changed for Huxley, but it continued to occupy his thoughts, fascinating him variously as a utopian ideal and as a lived experience and uniquely shaping his vision of the scientific history project that he launched under the auspices of UNESCO.

Complex life histories are good places from which to tell a story of grand visions and ideologically driven programs. The anthropologist Michael Fischer called such complex life stories of individual scientists *social hieroglyphs*, to distinguish them from biographies as such. Social hieroglyphs are incomplete in biographical detail or as accounts of individual career achievements, but "they provide historical cross sections to what are often told as only individual, institutional, or science or technical histories."[38] I use complex life stories in a similar way, as entry points for rearranging and supplementing the intellectual history of historiographic thought by recovering transnational, West-East circulations, interactions, and entanglements while revealing the individual strategies and local maneuvering that played out differently across global, national, and institutional spheres. By following scientists and scholars from Moscow or Leningrad to London or Paris or Washington, DC, or Philadelphia, and back to Russia, and back out again, this book reconnects the twentieth-century history of historical methodology—a backdrop against which the debates about bigger and deeper history are unfolding today—to the history of the radical ambitions and the changing meanings of socialism that variously animated people's imaginations for much of the twentieth century.

1 The Quest for Scientific History

Throughout the twentieth century, the notion of scientific history evoked strong reactions among professional historians. In 1903, when John Bagnell Bury, a newly minted regius professor at Cambridge, the highest among high-status professorships in the history profession in England, used his inaugural lecture to champion history as "science, no less and no more," the rebuttal from his fellow historians was quick and harsh: what nonsense.[1] By the final quarter of the nineteenth century, the distinction between the humanities and the natural sciences had settled into the form that largely holds today. As part of the interpretative humanities, history was understood as distinct from the analytic sciences. Culture was distinct from nature, the objective distinct from the subjective, and the interpretative distinct from the analytic. Even *Punch*, that venerable magazine of humor and satire, hammered it out, ridiculing scientific history in verse (see fig. 1).[2]

Yet, time and again throughout the twentieth century, a handful of historians championed the cause of scientific history. In 1929, when Bury's sharpest critic was named the new holder of the prestigious title, two historians at the University of Strasbourg, Lucien Febvre and Marc Bloch, inaugurated the *Annales d'histoire économique et sociale*, a journal that crystallized a model of history that became known as the *Annales* school of historiography.

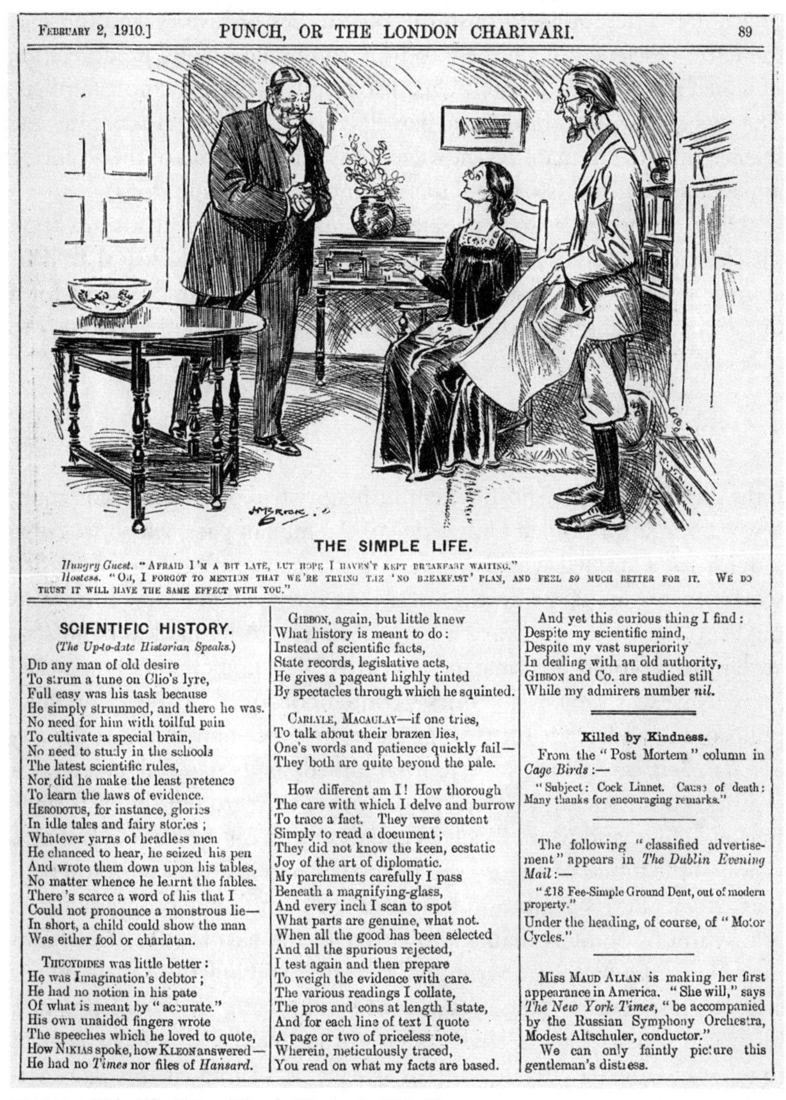

FEBRUARY 2, 1910.] PUNCH, OR THE LONDON CHARIVARI. 89

THE SIMPLE LIFE.

Hungry Guest. "AFRAID I'M A BIT LATE, BUT HOPE I HAVEN'T KEPT BREAKFAST WAITING."
Hostess. "OH, I FORGOT TO MENTION THAT WE'RE TRYING THE 'NO BREAKFAST' PLAN, AND FEEL *SO* MUCH BETTER FOR IT. WE DO TRUST IT WILL HAVE THE SAME EFFECT WITH YOU."

SCIENTIFIC HISTORY.

(*The Up-to-date Historian Speaks.*)

DID any man of old desire
To strum a tune on Clio's lyre,
Full easy was his task because
He simply strummed, and there he was.
No need for him with toilful pain
To cultivate a special brain,
No need to study in the schools
The latest scientific rules,
Nor did he make the least pretence
To learn the laws of evidence.
HERODOTUS, for instance, glories
In idle tales and fairy stories;
Whatever yarns of headless men
He chanced to hear, he seized his pen
And wrote them down upon his tables,
No matter whence he learnt the fables.
There's scarce a word of his that I
Could not pronounce a monstrous lie—
In short, a child could show the man
Was either fool or charlatan.

THUCYDIDES was little better:
He was Imagination's debtor;
He had no notion in his pate
Of what is meant by "accurate."
His own unaided fingers wrote
The speeches which he loved to quote,
How NIKIAS spoke, how KLEON answered—
He had no *Times* nor files of *Hansard*.

GIBBON, again, but little knew
What history is meant to do:
Instead of scientific facts,
State records, legislative acts,
He gives a pageant highly tinted
By spectacles through which he squinted.

CARLYLE, MACAULAY—if one tries,
To talk about their brazen lies,
One's words and patience quickly fail—
They both are quite beyond the pale.

How different am I! How thorough
The care with which I delve and burrow
To trace a fact. They were content
Simply to read a document;
They did not know the keen, ecstatic
Joy of the art of diplomatic.
My parchments carefully I pass
Beneath a magnifying-glass,
And every inch I scan to spot
What parts are genuine, what not.
When all the good has been selected
And all the spurious rejected,
I test again and then prepare
To weigh the evidence with care.
The various readings I collate,
The pros and cons at length I state,
And for each line of text I quote
A page or two of priceless note,
Wherein, meticulously traced,
You read on what my facts are based.

And yet this curious thing I find:
Despite my scientific mind,
Despite my vast superiority
In dealing with an old authority,
GIBBON and Co. are studied still
While my admirers number *nil*.

Killed by Kindness.

From the "Post Mortem" column in *Cage Birds*:—

"Subject: Cock Linnet. Cause of death: Many thanks for encouraging remarks."

The following "classified advertisement" appears in *The Dublin Evening Mail*:—

"£18 Fee-Simple Ground Dent, out of modern property."

Under the heading, of course, of "Motor Cycles."

Miss MAUD ALLAN is making her first appearance in America. "She will," says *The New York Times*, "be accompanied by the Russian Symphony Orchestra, Modest Altschuler, conductor."

We can only faintly picture this gentleman's distress.

FIGURE 1 "Scientific History," *Punch*, February 2, 1910, 89.

Regardless of their differences, the historians associated with the *Annales* school, from Febvre to Furet, all cast their craft as *scientific history*, seeking to reconcile biology, geography, and climate, on the one hand, with "political and cultural expressions of specific groups of individuals," on the other, as different agents of human history.[3]

Today, a number of historians and scientists are seeking new common ground. Whether they succeed or not, the most recent generation of scholars attempting to push the boundaries of history and those of the natural sciences is unlikely to look too much like its predecessors, of which the *Annales* school of historiography is only the most familiar. The repeated calls to bridge the perceived or real gap that separates the sciences and the humanities, however, raise the question of the enduring appeal of scientific history and the semantics of contempt for it.

How might the history of scientific history look if historians took the enterprise seriously? In what follows, I attempt to do just that. This chapter provides a necessary background by examining the quest for a unified historical knowledge of the past and the unified methodology to investigate it as it manifested itself in the decades around 1900.

Two Unity of Science Movements

If the different proponents of scientific history had one thing in common, it was a quest for unified knowledge of the human past, which was also a quest for a unified worldview as a necessary condition for a unified world. In his inaugural lecture of 1903, Bury claimed that historians ought to be working toward a *unified* history of humankind and that such history would yet come to be written.[4] The unification of historical knowledge was the key for Bury's contemporary Henri Berr, a French philosopher and academic entrepreneur. In 1900, Berr founded the *Revue de synthèse historique*, a journal aimed, as its name signaled, at a synthesis of historical knowledge. In 1907, he recruited the historians Lucien Febvre and Marc Bloch as collaborators on the *Revue*. Febvre's vision of the future of history was shaped by his close collaboration with Berr. It was Berr's aspiration to instigate an ambitious program aimed at a synthesis of all available knowledge of the past that Febvre championed as *total history*, a term that became a shibboleth of the *Annales* school.[5] Insisting on "the necessity of synthesizing all knowledge in a historical framework," Febvre asserted that history should be central to what he called a "living Unity of Science" ("l'Unité vivante de la Science"), setting in motion an "organized and concerted group inquiry" into a unified "science of the human past" ("science du passé humain").[6]

The promotion of the unity of science is better known from a different movement, one launched by the philosophers and scientists of the so-called Vienna Circle.[7] In 1929—the year Febvre and Bloch founded the *Annales* with the goal of "uniting researchers of different origin and specialization"[8]—the Vienna Circle issued a manifesto announcing its own "collaborative effort" aimed at the unification of "different branches of science."[9] The Vienna Circle, which included the philosophers Moritz Schlick and Rudolf Carnap, the economist Otto Neurath, the physicist Philipp Frank, and the mathematician Kurt Gödel, among others, sought to systematize a formalized "logical structure of the world" (as Carnap put it) as a foundation for an antimetaphysical, "scientific world-conception."[10] With its explicit promotion of the unity of science in such endeavors as the international congresses for the unity of science and the *International Encyclopedia of Unified Science*, the Vienna Circle became the most common reference for the efforts of scientific unification in the twentieth century.[11]

One discipline, however, was notably absent from the Vienna Circle's unity of science project: history. The philosophers of the Vienna Circle rationalized that exclusion by distinguishing between what Hans Reichenbach, one of the leaders of the group, defined as "the context of justification" and "the context of discovery," in other words, between epistemological questions of justification of knowledge (i.e., How do we know what we know?) and the circumstances of how scientists supposedly discovered that knowledge. The Vienna Circle philosophers sought to develop a new, scientific philosophy and were primarily concerned with the context of justification of scientific knowledge. History, a discipline concerned with establishing the "truth" about the past, could reveal the unique circumstances of the discovery of scientific "truths," but, since the historical circumstances were unique, they had no epistemic consequences. Members of the Vienna Circle therefore saw history as a source of raw data for theory construction but otherwise irrelevant for the unity of science project.[12]

In distinguishing between the context of justification and the context of discovery, Reichenbach assimilated a well-established argumentative strategy that presented the relationship between the sciences (exact and objective) and the humanities (interpretative and subjective) as mutually exclusive forms of knowledge. This binary was articulated most forcefully in the Germanic intellectual tradition in the third quarter of the nineteenth century. In his influential *Introduction to the Human Sciences*, first published in 1883, the Berlin philosopher Wilhelm Dilthey, for example, has argued that the natural sciences and the humanities had different objects of study.[13] The object of the humanities, he stated, was the human mind and spirit (*Geist*), as expressed in the historical process. In contrast to the objects of the natural world, these objects are invested with meaning. Unlike the material world of scientists, the world that historians study is the product of human labor and a creation of human interaction. Causal explanations of the natural sciences, Dilthey argued, could never account for human action or capture a historical world made of inherited meanings. Consequently, the task of the humanities (*Geisteswissenschaften*) is not causal explanation (*Erklären*) but rather understanding (*Verstehen*) and interpretation (*Deutung*).[14] In history, according to Dilthey, understanding is possible because historians—unlike scientists—could "live through" (*Erleben*) the experiences of other humans and comprehend their actions. In other words, unlike the natural sciences, history is hermeneutical in its nature.[15]

The leading members of the Vienna Circle, including Rudolf Carnap, Moritz Schlick, and Otto Neurath, read Dilthey and assimilated his line of reasoning in their own critique of metaphysics. Carnap, in particular, credited Dilthey's critical hermeneutic in his own work, stressing that "the intuitive feeling of life" cannot be part of the unity of science system because it cannot be conceptually determined.[16]

Although actual practices in the humanities and the sciences, both now and in the late nineteenth century, defy a simple binary opposition between understanding and explaining or between general and unique, a view of the relationship between the humanities and the sciences as mutually exclusive forms of knowledge was perpetuated in the writings of such influential philosophers as Heinrich Rickert, Ernst Cassirer, and Hans-Georg Gadamer.[17] Within the philosophy of science, Reichenbach's distinction between the context of discovery and the context of justification reigned for most of the twentieth century.[18] The binary between the sciences and the humanities was prominently echoed in the two cultures dichotomy that C. P. Snow postulated in the 1950s, and it continues to powerfully reverberate today.[19]

A very different movement for the unity of science had been unfolding in France since the last decade of the nineteenth century. In the French context, the talk of unity was anchored in a philosophical tradition with roots in the writings of Auguste Comte and other French representatives of positivist thought. The Vienna Circle's insiders and their later interpreters used the language of *neopositivism* or *logical positivism*, but the Francophone scholars applied the term *positivist* to different questions and problems than did their Germanophone peers.[20] For the French positivists, history was deemed to be at the center of the unified system of knowledge envisioned by Comte and variously reconstituted by his later readers, followers, and opponents.[21] It was this all-embracing concept of history rooted in the discussions of Comtean unity that underpinned Henri Berr's historical synthesis and set the stage for the *Annales* project.

Born in 1798 in Montpellier, Auguste Comte (1798–1857) devoted his life to developing and promoting a *philosophie positive*—a "new unified system of knowledge for the modern, industrial era."[22] His quest ultimately harked back to the French Revolution. At a young age, he was deeply moved by the Revolution's attempt to replace a theologically grounded monarchy with a secular society based on science. In 1816, after he was expelled from the École polytechnique for insubordination, he went to work for the social theorist and reformer Henri de Saint-Simon, a visionary proponent of constructing a new unified system of scientific knowledge, or positive philosophy, that would, he argued, lead to a new stage of history in which scientists and industrialists would replace priests and military leaders. Comte would ultimately part ways with Saint-Simon to embark on his own lifelong journey to develop these ideas into a coherent system, all the while battling mental illness and trying, unsuccessfully, to obtain a professorial position. From 1830 to 1842, earning his living as a teaching assistant and an admissions officer at the École polytechnique, he wrote his six-volume magnum opus, *Cours de philosophie positive*, an encyclopedic outline of the system of positive philosophy.[23]

Comte's system was based on an understanding of society as a reflection of the dominant philosophy of nature at the time. In the *Cours*, Comte presented an epistemic merger between nature and society through two key conceptual elements, the famous law of three stages and the sixfold classification of the sciences. The former stated, as Comte put it, that "each of our principal conceptions, each branch of our knowledge, passes in succession through three different theoretical states" or "three methods of philosophizing"—the theological, the metaphysical, and the positive—that are "radically opposed to each other."[24] The law of three stages was vindicated, according to Comte, by the actual historical development of the six fundamental sciences—mathematics, astronomy, physics, chemistry, biology and "social physics" (or *sociology*, Comte's neologism for a "science of society").

Since, according to Comte, society reflected the dominant philosophy, it passed through the same states in the same successive stages. In the theological stage, it was dominated by ideas of divine right and, therefore, ruled by priests and military leaders. In the metaphysical stage, the "abstract and metaphysical assumption of an original social contract" is preeminent, and "rights, viewed as natural and common to all men to the same degree," are guaranteed by the power of lawyers and the judiciary. In the final, "positive" stage, in which each branch of science would achieve its positive state, absolute notions would be replaced with relative ones, all people would agree on basic notions and ideas, and society would regain stability by becoming a republic of knowledge ruled by positive philosophers.[25]

In the positive stage, which was still to come, the whole of knowledge would be unified in the sense that each science would submit to the same methodological standard. Comte emphasized that the main value of science lie in its method. Each science in his sixfold hierarchy contributed something unique to the "positive method": mathematics gave the method of analysis, astronomy gave observation and hypothesis, physics gave experimentation, biology gave comparative method, and sociology gave a "synthetic" historical method (chemistry was an exception to this rule since, according to Comte, it did not add any new aspects to the experimental method of physics). In this succession, each science built on its predecessor. As each passed through the stages—from analysis, to observation, to experiment, to comparison, and, finally, to the historical method—its methods gained new aspects. While scientific doctrines and theories changed over time, the methods of the different disciplines drove the development of scientific knowledge. The historical method, the last one in Comte's hierarchy, was understood not just as a method of sociology but as a synthetic method that would permeate all sciences in the positive stage.

Comte's *Cours* and his subsequent lectures on the subject attracted significant attention during his lifetime.[26] Comte was still not able to secure a desired job as a professor, but his lectures were well attended, and his writings gained wide audiences. In Britain, his admirers included the philosopher John Stuart Mill, who helped popularize his teaching across the Channel. In France, the noted scholar Émile Littré became a disciple and an early champion of Comte's teaching.[27] He also organized a "positivist subsidy," a fund that provided for Comte in the final years of his life.

As Comte grew older, he became more conservative, to the dismay of his republican followers such as Littré. After the revolutions of 1848–49, Comte praised Napoléon III and celebrated the restoration of the empire. He wrote disapprovingly of revolutionary doctrine, positioning his own doctrine, with its law of social development and progress, as a healthy alternative.[28] He also started to promote what he called a *religion of humanity* (a term he borrowed from Saint-Simon and his followers), a view of the positivist philosophy as a universal secular system of beliefs that would ultimately replace Catholicism. In his last years, he developed these ideas in his most controversial works, *System of Positive Polity; or, Treatise on Sociology: Instituting the Religion of Humanity* and *The Catechism of Positive Religion*.[29] These ideas appealed to skeptics, especially in England, where a group of Oxford students established the London Positivist Society and the London Church of Humanity.[30]

In Germany, Comte's positive philosophy had a more complicated reception. His teachings reached Germany in the 1850s. The German mandarins debated his ideas in the context of ongoing Germanophone debates following the revolutions of 1848–49 that used the language of positivity in quite a different sense than the French did. Conflating ideas about positivity as a defining feature of the natural sciences with laments about what they perceived as the assault of the natural sciences on the German "historical-philosophical manner of education [*Bildungsweise*]," in their defense of traditional forms of political and academic authority they decried positivism as a major threat to sound scholarship and a national sense of community.[31] When, half a century later, the members of the Vienna Circle, who developed their work largely independently of Comte and his French legatees, championed their program under the banner of neopositivism, they sought to appropriate the radical status assigned to positivism in the Germanic academic world rather than its intellectual content.[32]

If in Germany *Positivismus* became a derogatory label among intellectuals and politicians alike, in France Comte's influence rose dramatically during the early years of the Third Republic following that country's defeat in the Franco-Prussian War of 1870, which ended the second Napoleonic empire. In historiography, it became customary to describe the Third Republic as *positivist* in its ideology.[33] Leading politicians of the Third Republic, such as Jules Ferry, described themselves as *disciples* of Comte and embraced positive philosophy. Hoping to replace the authority of the Catholic Church in public life with "positive science," these leaders repackaged Comte's doctrine, associating it with their anticlerical and educational policies.[34] Émile Littré, in particular, helped rebrand Comte's teaching as an exciting worldview dedicated to the cult of science. As a member of the republic's political elite, first as a parliament deputy (1871–75) and then as a lifetime senator (1875–81), he claimed that Comte's positive philosophy, which linked "mental and social stability to the stability of science," represented "the remedy for a troubled era."[35]

As a quasi-official philosophy of the early Third Republic, Comtean positivism was reconciled with various programs and predispositions in areas ranging from literature to the workers' movement.[36] Positivism's appeal rested not only on its strong sociopolitical promise but also on the malleability of its content as a framework for the unification of knowledge. History—which was institutionalized in the French academy during this tumultuous period—became one of the "laboratories of positivism" in which Comtean positive philosophy was politically repackaged and conceptually refurbished.[37]

Positivism, History, and Henri Berr's Historical Synthesis

In France, historical studies received a major boost in the years between the so-called second and third French revolutions of 1830 and 1848.[38] Regarded as being of little use to the state during and after 1789, history gained a renewed purchase as an instrument to heal the divisions in the French nation that stemmed from the Revolution, and, at the same time, to justify the Revolution as the foundation of a new era, one in which conflicts and contradictions faded.[39] In this period, the state founded and fully funded a number of historical institutions, such as the Committee for Historical Works and the Commission for Historical Monuments, that employed a growing number of historians.

Professional historians in France, as elsewhere in Europe, adopted the standards of history writing based on archival research and careful criticism of documents that had been established by Leopold von Ranke. The Rankean method of historical training then spread from the University of Berlin, where Ranke taught from 1824 to 1871, throughout Europe and North America.[40] In the 1860s, and especially in the early years of the Third Republic, when positivism entered the debates of politicians and academics of different stripes and resurfaced as the dominant sociopolitical philosophy in France, the historians blended Comtean positivism, by then reevaluated by Littré and dissociated from its author, with their established practices of history and the historical method.

The expression *positivist history*, or *the positivist school* of historiography, became a common label attached to the group of Parisian historians known as the *école méthodique*, which included Charles-Victor Langlois, Charles Seignobos, Numa Denis Fustel de Coulanges, and Gabriel Monod, among others. Many of *méthodiques*' premises, however, such as the fondness for the document, were nourished by Ranke rather than Comte.[41] As Fustel de Coulanges wrote in 1888: "The best historian is the one who keeps close to the texts, who interprets them most accurately, who thinks and writes only according to them." According to Fustel: "History is not an art; it is a pure science."[42] By *science* he meant a methodical analysis and critique of documents to ensure that their testimony was valid. In the first book about historical method published in France, the 1898 *Introduction aux études historiques*, Seignobos and Langlois stressed: "History is made with documents. . . . Without documents, there is no history."[43]

During the early years of the Third Republic, when the budget for higher education tripled and history, now regarded as a discipline at the forefront of the human sciences, became the focus of great investment and enthusiasm, the historians associated with the *école méthodique* formed the intellectual vanguard of the Parisian academic community. By the close of the nineteenth century, however, the *méthodiques* began to show signs of crisis. As the historian Isabel Noronha-DiVanna noted in her study of the group, their efforts to professionalize history and refine its method inadvertently separated their field from the general readership on which their position as state-funded employees depended.[44] They also made themselves a target for the pens of their adversaries by transgressing the boundaries of their discipline and making larger claims for the role of historical method, in a vaguely Comtean fashion. *La méthode historique appliquée aux sciences sociales* (the new edition of the *Introduction aux études historiques*), for instance, featured sections that discussed the application of the historical method to all aspects of social reality, positioning historical method as a true method of sociology.[45] Not surprisingly, it was the representatives of the new academic field of sociology, which became institutionalized in France in the 1890s, who were the most vocal critics of the *école méthodique*.[46]

Henri Berr's *Revue de synthèse historique* became the main venue of the organized attack on the *école méthodique*. In his inaugural editorial, Berr stated that the *méthodiques* as well as historians more generally inadequately understood the historical method. Despite the "renaissance of historical scholarship in France after 1870," he stressed, historical studies remained primarily in the stage of "historical analysis." Against the approach that he called, pejoratively, *histoire historisante*, he sought to use the new journal to provide a space in which "the power of analysis would be combined with the spirit of synthesis," hence the *synthèse historique* in the name of the journal.[47] Using Seignobos and Langlois as a foil, Berr contrasted his program of historical synthesis to the "erudite synthesis" of the historians. His was to be a "scientific" synthesis. As he stressed in his 1911 *Synthesis and History*: "Historical synthesis must be science, true and complete science. This is precisely what is meant by the word *synthesis*."[48]

Berr usually appears in the historiography as a principled critic of "positivist history."[49] Berr's pejorative *histoire historisante* became synonymous with *positivistic history* after Lucien Febvre and other Annalistes built their own identity by attacking the older generation of Seignobos and Langlois as examples of "history that is not ours."[50] But, while Berr distanced himself from Seignobos and other mainstream historians, he presented his own program in an explicitly Comtean fashion, even as he decisively modified and refurbished its content. His program of historical synthesis, and his efforts to implement it in practice, in turn, laid the groundwork for much of the subsequent development in French historiography around the *Annales* school.

Berr first presented his historical synthesis, or the "science of history," in an article published in a literary magazine, *La nouvelle revue*, in 1890. The article was permeated with the Comtean language of *stages* and Comte's sixfold hierarchy of sciences. In the opening paragraphs, Berr stated: "A new science is about to be born, one that has been long awaited, often promised, but always deferred; of all the sciences, this is the last to appear but the first in importance, and one that builds on most of the others and surpasses them, for its subject matter lies in the most complex of phenomena: those that constitute the evolution of humanity." This new science would make the new, scientific, "synthetic stage" possible. In the synthetic stage, history was to become nothing less than a science of the "total reality" of the unity of humanity. "If we are to adopt the three phases of Auguste Comte," Berr wrote, "[then we may well expect] the 'new science' to finally arrive at the positivist phase."[51]

From 1894 to 1899, while pursuing his doctoral studies at the prestigious École normale supérieure, Berr developed an ambitious program of historical synthesis that he presented as a polemic with neo-Kantian philosophical tradition. He argued that the neo-Kantian philosophy, which emphasized the deductive method and the notion of immutable a priori systems of knowledge, reached a stalemate. In his doctoral thesis, ambitiously titled "The Future of Philosophy: An Outline of a History-Based Synthesis of Knowledge," he argued that natural laws are not static or fixed but change over time. As the only discipline dealing directly with change over time, history ought to be a unifying framework for all knowledge.[52]

In this polemic addressed to philosophers, Berr found an ally in the mathematician and philosopher of science Henri Poincaré. Berr's writings are peppered with references to Poincaré, in particular to his 1911 paper "The Evolution of [Natural] Laws," in which Poincaré discussed the possibility that laws might change over time.[53] Berr invoked Poincaré's conception of geometric conventions as hybrid analytic-synthetic expressions "guided" by experience to argue for a new "logic of history." History operated neither by Newtonian "natural laws" nor by Kantian "a priori" models, he argued. Instead: "It is necessary to work on establishing a [new] logic of history, substituting the concept of a priori with inductive logic born out of experience, and [making this] a routine practice."[54]

Endorsing a Comtean view of history as a central methodology of all the human sciences, Berr prioritized the cooperative nature of knowledge. He championed his grand idea of historical synthesis as something that would be achieved through collaboration between specialists from different fields. The notion of synthesis as a collaborative effort not only underpinned Berr's writings but also materialized in his efforts to create the institutions to carry on his program.

In his editorial inaugurating the *Revue de synthèse historique*, Berr described the purpose of the journal as "an appeal for greater cooperation between scientists and historians": "Historical synthesis is intended to induce various teams [to work] together, . . . to become of greater mutual assistance as the conception of the common goal becomes clearer." The journal invited contributions from scholars and scientists from various disciplines to "complete" the program of various "scientific syntheses" that already had begun and to bring about a genuine science of history (*histoire-science*).[55]

Over time, the *Revue* featured articles by historians, psychologists, sociologists, and the natural scientists.[56] Lucien Febvre, who first published in the *Revue* when he was a young student at the École normale, would become Berr's closest collaborator and an ally. As he recollected in 1925: "We were a group of young historians at the *École normale* who were beginning to find our studies banal, and were just about to quit when in 1900, our interest in history was re-fired by the appearance of the *Revue de synthèse historique*."[57]

In 1920, Berr launched an encyclopedic series, *L'évolution de l'humanité*, a collaborative project aimed at producing a "universal history." In the preface to the first volume, he described the project as follows: "The enterprise is a vast experiment to unfold bit by bit before the eyes of the public and for the great profit of historical science, [the experiment] in which the ideas put forward for testing will be confirmed or corrected. . . . Each [volume] will be an individual oeuvre, carrying the mark of a personality."[58] Febvre's volume for the series, *La terre et l'évolution humaine* (1922), became an emblematic work of the *Annales* school.[59]

In 1925, Berr founded the International Center of Synthesis.[60] One unit of the center, the Section of Historical Synthesis (Section de synthèse historique), was entrusted to Febvre, who codirected it with his patron. Another unit, the Section of the Natural Sciences (Section des sciences de la nature), was directed first by the philosopher Abel Rey and then by the physicist Paul Langevin. Within few years, the center grew into a veritable institution outside the French academy, enjoying a large and distinguished membership. Its administrative council listed such scientific celebrities as Albert Einstein, Ernest Rutherford, and Émile Roux as well as the directors of the leading French scientific institutions, such as the École normale supérieure and the Muséum d'histoire naturelle.[61]

Collective work was the defining feature of the center. Its activities were organized around annual events, "the weeks of synthesis" (*les semaines de synthèse*).[62] Each year between 1929 and 1939, Berr and the council would select two themes, one from the natural sciences, one from the humanities. During the week of synthesis, speakers representing the *lettres* (the humanities) and the natural sciences would discuss both themes, attending each other's sessions. In 1929 the themes of the week were evolution and civilization, in 1930 relativity and the origins of society, in 1931 quantum theory and individuality, and so forth. According to the historian Bernadette Bensaude-Vincent, who analyzed the published proceedings of the weeks of synthesis over several decades, these exclusive, invitation-only events brought together astronomers, mathematicians, physicists, and biologists along with philosophers, historians, anthropologists, linguists, and sociologists.[63]

The weeks of synthesis did not yield a synthetic account of knowledge unified through history (the program of historical synthesis as Berr outlined it in his published work). But this was not their purpose. Instead, Berr and other leaders of the center emphasized programming and discussions. An exploratory seminar rather than a synthesizing publication was the main objective of the weeks. Invited experts provided overviews of the state of knowledge on designated themes, and participants explored the topic in discussion sessions.[64] As a testing ground for cross-disciplinary conversations, the center, the series *L'évolution de l'humanité*, and the *Revue* proved to be very successful ventures that provided a forum for discussing and refining Berr's program of historical synthesis, especially among the younger generation of historians who later formed the *Annales* school of history. Febvre emphasized that, just as he did, the *Annales* school's leaders endorsed Berr's view of history as a discipline that "must be wide open to the findings and methods of other disciplines . . . and at the same time must resist the temptation . . . to divide itself into a number of 'specialisms' . . . each going its own independent way."[65]

The first week of synthesis, which inaugurated the center in the summer of 1929, also launched two other institutions. The first was the International Committee of the History of Science, which was formed at the international historical congress in Oslo a few months earlier but met for its first organizational meeting as part of the week of synthesis. The second was related to the first, stemming from the committee's decision to inaugurate a regular international history of science congress: the organizational meeting of the Committee, held at the International Center of Synthesis in conjunction with the first week of synthesis, was rechristened the First International Congress of the History of Science.[66]

The beginnings of international organizations for the history of science illustrate how a minor field that had shown only limited signs of institutionalization before 1900 received a major boost from the quest for the unity of knowledge rooted in Comtean positive philosophy. In Berr's rendering, the role of the history of science in the positive stage of historical synthesis was to serve as a liaison between the sciences of natural phenomena and those of human life and between the historians and the scientists.

Historical Synthesis and the History of Science

The launch of the *Revue de synthèse historique* was timed strategically to coincide with a major international event, the Paris Exposition of 1900 (Exposition universelle internationale de 1900). The fifth world's fair to be held in Paris and the biggest since 1855, the 1900 exposition was a far-reaching event, featuring exhibitions that showcased achievements of different nations and even staging the second Olympic Games. Over fifty million visitors and eighty-three thousand exhibitors flocked to the French capital from all corners of the globe.[67] The exposition offered a major showcase for all things French—technical and scientific achievements, industry, economy, and culture. More than a hundred academic congresses were organized in conjunction with the festivities.[68] Since the later part of the nineteenth century, when international congresses became a familiar part of academic life, scientists and scholars had used these gatherings to highlight the significance of their discipline and project a vision for its future. Like other academics, historians used the 1900 exposition as an opportunity to promote their discipline through the venue of the international congress, gathering in Paris for the First International Historical Congress.[69]

Berr, a participant in the congress, used the occasion to give his program of historical synthesis wider publicity.[70] In his presentation at the congress, he positioned historical synthesis as part of the relatively new field the history of science. He registered at the congress as a "historian of science," and he spoke of his program at the congress's session on the history of science.[71] The fact that the congress featured a session on the history of science at all is itself a testament to the status of this relatively new field in France, a status owing in large part to the role assigned to the history of science within the Comtean system of thought.

Describing scientific knowledge as a social and historical product, Comte emphasized that "the true philosophy of every science is necessarily inseparable from its actual history."[72] Historians have credited him with initiating a historically informed philosophy of science as a distinct discipline in France.[73] He also played a major part in the institutionalization of the history of science as a separate field in France. In 1832, he tried, unsuccessfully, to create a chair in the history of science at the Collège de France. While his own attempt failed, his disciples, a teacher of mathematics, Pierre Lafitte, and a Russian count residing in Paris, Grégoire Wyrouboff, later succeeded in establishing such a chair under the regime that favored Comtean doctrine. In 1892, the Chair in General History of Science was created at the Collège de France, with Lafitte as its first occupant.[74] By 1900, the history of science had achieved an unparalleled status in France, with Paris as an important center for a field that otherwise had barely begun to show signs of institutionalization.

At the historians' congress, the history of science was given pride of place as one of the five themes of the congress.[75] Devoting a separate symposium to the history of science, the organizers not only recognized that the field had a strong footing in the French academy but also used the history of science session to showcase the Comtean theme, which reverberated powerfully throughout the exposition. Opening the congress in the midst of festivities all around Paris, the congress's president, Gaston Boissier, started with the motto of the exposition—"to define the philosophy and express the synthesis of the nineteenth century"—declaring that the century should be called "the century of history." Historical method, he stated, had triumphed across the disciplines, facilitating contacts between the different sciences, and easing mutual understanding between different nations.[76] Another of the congress's organizers, the Paris-based professor of legal history and constitutional law Adhémar Esmein, continued the theme, stating that, at the threshold of the new century, history had become a *science maîtresse*—one of the "leading sciences"—on par with the natural sciences and technology.[77]

The imprint of the Comtean characterization of the role of history and its unique contribution to the positive method is evident in the design of the congress. It gathered under the theme of comparative history and highlighted, in Comtean fashion, comparison as the main method of scientific history.[78] In the Comtean sixfold classification of the sciences, biology introduced the method of comparison, but history was to become the ultimate manifestation of the comparative method in the positive stage.[79] In keeping with the organizers' agenda, Berr spoke of his program of historical synthesis as the embodiment of comparative methodology.[80]

Discussion of Comte featured prominently at the history of science session of the congress. Of the dozen presentations published in the proceedings, two were devoted entirely to discussion of Comte's thought and its legacy.[81] One of these was given by Gaston Milhaud, a professor of the history of philosophy at the University of Montpelier, the other by Eugène Gley, a professor of medicine at the Collège de France. While both speakers shared Comte's dynamic view of knowledge in the making, they called for major revisions and updates of Comtean positive philosophy. Milhaud took issue, in particular, with Comte's religion of humanity criticizing his later writings on the subject as an expression of his increasingly conservative attitudes as manifested in his penchant for Catholicism.[82] Gley, on the other hand, criticized Comte's endorsement of the notion of the fixity of species and his rejection of the "transformist" thought of Jean Baptiste Lamarck. In view of the advancement of the "biology of evolution" in the past decades, he called for the major refurbishing of the positivist framework.[83]

For Berr, the history of science was a vital component of his program of historical synthesis. In the years following the congress, Berr twice (in 1903 and 1906) submitted his candidacy for the chair in the history of science at the Collège de France. Despite the support of the celebrity philosopher Henri Bergson, he did not receive it.[84] Undeterred, he created his own institution for the history of science. When he outlined his vision of the International Center of Synthesis in 1928, he gave the history of science pride of place. As he explained: "There is nothing more important for the synthesis of knowledge, our main object, and for synthesis between speculation and action, the main goal of our efforts, than the history of science, as it finds itself in a position to establish a tight liaison between the sciences of nature and those of the humanities."[85]

At first, the history of science was part of the center's Section of the Natural Sciences, initially headed by the philosopher Abel Rey. After a reorganization in 1929, Berr invited Aldo Mieli to head the separate Section of the History of Science. An Italian intellectual and a passionate promoter of the history of science who had in 1919 founded a journal for the history of science, *Archivo di storia della scienza* (renamed *Archeion* in 1927), that he ran from Rome, Mieli had moved to France to escape Mussolini's secret police.[86] He was a driving force behind the efforts of the small international community of historians of science to promote their field through the venue of the international historical congress. In 1928, at the Sixth International Historical Congress in Oslo, he marshaled support to form an international committee on the history of science. At the International Center of Synthesis, he took on the administration of the nascent international committee. In the following years, he integrated the History of Science Section and the committee's work into the larger framework of Berr's programs, beginning with the first meeting of the International Committee of the History of Science, which was held during the first week of synthesis in the summer of 1929. As the published proceedings of the week noted, the timing of the history of science meeting was "intentionally coincidental" with the first week of synthesis, which, as we have seen, a mix of biologists, historians, anthropologists, and philosophers discussed, in parallel sessions, evolution and "civilization, the word and the idea."[87] Some of the participants in the week of synthesis, such as the biologist Maurice Caullery, also participated in the meetings of historians of science.[88]

The setting and the agenda of the First International Congress of the History of Science left an imprint on the planning of the second. Like the week of synthesis, which inaugurated the first congress, the second congress, held in London in 1931, would feature parallel sessions on themes relevant to the sciences and the humanities. Berr's positioning of the history of science as part of an ambitious program of historical synthesis also invested the seemingly arcane field with larger significance. As we shall see in the next chapter, the London congress would become the stage on which an alternative vision of historical synthesis was presented by the Soviet delegation, which was headed by the Bolshevik politician and Marxist scholar Nikolai Bukharin.

Like Comte and his disciples, who instituted the history of science in the French academy, the successive promoters of this nascent field had grand aspirations for their scholarly programs. In the colorful prose of the historian of science George Sarton, a corresponding member of the International Center of Synthesis who was strongly attracted to Comte's teaching: "The history of science is the history of mankind's unity, of its sublime purpose, of its gradual redemptions."[89] Sarton founded what today remains the journal most associated with the history of the science, *Isis*, in 1913. Having fled Belgium after the outbreak of World War I, first to England and then to the United States, he initially worked for the Carnegie Foundation for International Peace before obtaining a research position at the Carnegie Institution in Washington, which subsidized the publication of *Isis* for several years after the end of the war in Europe.[90] After Sarton finally settled as a lecturer at Harvard, where he developed the first curriculum in the history of science, he would promote the history of science as the ultimate expression of what he called the *new humanism*—a holistic, all-embracing vision and a worldview based on the appreciation of science in history as a means to achieve worldwide unity and solidarity.[91] As he declared in *The History of Science and the New Humanism*: "The unity of science and the unity of mankind are but two aspects of the same truth."[92] In a fashion quite Comtean, Sarton and other practitioners of the history of science in the first tumultuous decades of the twentieth century were promoting a worldview. Their worldview, unlike that of Comte, however, was distinctly early twentieth-century internationalist and pacifist. The history of science was the means to an end, where the end was the world peace.

The Internationalist Politics of Synthesis

Like many other promoters of the unity of knowledge in the first decades of the twentieth century, Sarton reconciled his scholarly preoccupations with political activism as an *internationalist*. In 1913, the year he founded *Isis*, he contributed a series of articles to *La vie internationale*, the programmatic internationalist periodical funded by the Carnegie Endowment for International Peace and published in Brussels by the Union of International Associations. *La vie internationale* was the brainchild of the socialist and pacifist Henri-Marie Lafontaine, the union's cofounder and its secretary-general, who was awarded the Nobel Prize for Peace just before the outbreak of World War I. In his articles for Lafontaine, Sarton discussed in the same breath the role played by a new kind of professional like him, the *organisateur internationaux*, in agitating for internationalism and the role of the history of science as a vehicle for the "demonstration of internationalism . . . in action."[93] The history of science was an instructive resource for students of international affairs, he argued, because "science is the world's communal thought" and scientists' loyalties were to communities of thought rather than to particular nations.[94]

Sarton and his circle of internationalist friends foreshadowed the ascent of a distinct kind of internationalist movement that the historian Geert Somsen has aptly called *socialist internationalism*.[95] Like many other members of his generation, Sarton was attracted to socialism. As a student in Ghent, he studied Karl Marx, but he rejected Marx's revolutionary teaching in favor of Fabian socialism, which emphasized the peaceful and gradual transformation of society from within.[96] He was actively involved in the student socialist movement. In 1910, he helped found the Society of Socialist Students in Ghent and was its president for two years. When he founded the journal *Isis* in 1913, Sarton saw it as a medium of internationalism and socialism. It was only in the 1920s, when he reestablished himself in the United States, that he reframed *Isis*'s philosophy in apolitical terms. As his biographer put it: "*Isis* paid less attention to socialism than to pacifism, which [Sarton] often invoked in the form of promoting human kindness."[97] Sarton's personal views on socialism, however, did not change. When, in 1926, an old acquaintance from his student days in Ghent, the socialist leader Hendrick De Man, sent him the manuscript of his forthcoming *The Psychology of Socialism*, Sarton responded with enthusiasm and wrote a sympathetic review.[98]

This form of internationalism, which gained prominence between the two world wars, was distinct from the earlier ethos of internationalization in that it endorsed a vision of unified science as an antidote to nationalistic sentiments and militarism, a vehicle to advance the cause of peace, and a model for a unified world government.[99] A derivative of the adjective *international* (coined by the British philosopher Jeremy Bentham in the late eighteenth century in the context of "international law"), the term *internationalism* entered English-language dictionaries at the end of the nineteenth century connoting a distinct set of values, meanings, and politics. It was originally used in the context of the doctrine of the International Workingmen's Association, a socialist organization formed in London in 1864 under the leadership of Karl Marx.[100] At the turn of the century, dictionaries defined *internationalism* as cooperation between governments or their citizens aimed at material or moral betterment in the interest of the whole social order.[101] Internationalism was, in other words, a political movement that implied a new way of thinking about world affairs, less rooted in traditional concerns with international cooperation and national competition and more concerned with an ideal of a futuristic postnationalist, global, and cosmopolitan society.

As a political movement, internationalism coincided and overlapped with the internationalization of science. Both trends intensified in the decades around 1900, facilitated by new communication technologies, the increased number of translations, and the ease of travel.[102] For some scientists and scholars, internationalization was a means to an end. They capitalized on their international networks and used the tools of international organizations to gain the legitimacy they often lacked in their own countries.[103] Others, however, embraced the goal of internationalism as a political end. The champions of the historical synthesis discussed in this chapter belonged to the latter category.

When he founded the International Center of Synthesis, Berr meant it to be a distinctly *internationalist* institute, one that combined professional and political commitments. From the time of its official inauguration, each of its four units featured an equal number of French and foreign members.[104] The center was established with strong institutional connections to other internationalist organizations, including the League of Nations and its branches—the International Committee on Intellectual Cooperation, which operated from the league's headquarters in Geneva, and the International Institute of International Cooperation (IIIC) in Paris, an executive branch of the league. In 1925, a few months before the IIIC opened its grand premises in the Palais royal in the center of Paris, its planners approached Berr with a proposal to open a center on its premises.[105] Tensions over specific arrangements eventually derailed the negotiations, but the two organizations maintained a close connection throughout the 1920s and 1930s. The secretary of the IIIC, André Toledano, was also the secretary of the center.[106] The IIIC director, the former French government official Julien Luchaire, joined the center's administrative council.[107]

The two organizations—the political (the IIIC) and the intellectual (the International Center of Synthesis)—were not just close, they were pursuing the same project. The IIIC embraced the notion of the unity of knowledge, promoting it through a diverse range of programs.[108] At the League of Nations, the unity of knowledge was seen as a means toward the end of realizing its internationalist visions, promoted by intellectuals who were assigned the explicit role of diplomatic agents. The term *intellectuals* itself gained prominence at this time, denoting scientists and scholars who combined academic activities with public intervention.[109]

Like the League of Nations, the International Center of Synthesis had strong socialist associations. Funded by the French government, with government officials on its advisory board, it was directed by Berr but officially presided over by Paul Doumer, the radical elected speaker of the French Senate who became the president of France in 1931. Doumer remained the president of the center until his assassination in 1932. With its strong left-wing orientation, the center was described as a "child of the *cartel des gauches*."[110]

Politically, Berr was, like many French intellectuals of his generation, on the left.[111] A number of scholars and scientists associated with the center were socialists. Febvre, Berr's right hand at the center, was close to the Socialist Party and maintained contacts with the Radical Party and the Communists. Another of Berr's close associates, the physicist Paul Langevin, was a Marxist and a leading figure of the French scientific

Left who joined the Communist Party during World War II.[112] Langevin was also actively involved in the League of Nations as a member of its International Committee on Intellectual Cooperation.[113] He promoted an ideal of unified science as a precondition of a unified world under the aegis of both the League of Nations and the International Center of Synthesis.

For this generation of socialists, whatever the specifics of their political platforms, the Russian revolutions that culminated in the establishment of the first socialist state represented a key political frontier. These circumstances uniquely positioned Russian historians, both before and after 1917, in the international movement around scientific history.

2

Scientific History and the Russian Locale

The quest for unified knowledge, which gathered momentum in the late 1920s, coincided with the entrance of Soviet Marxist scholars into the international arena. They presented a competing, Marxist vision of unity. In 1928, a delegation led by the Bolshevik politician and Marxist historian Mikhail Pokrovskii attended the international historical congress in Oslo.[1] Five years later, another Bolshevik politician and the leading Soviet systematizer of Marxism, Nikolai Bukharin, arranged to debate Marxist synthesis with Henri Berr at the next international historical congress in Warsaw. In between, in the summer of 1931, Bukharin used the Second International Congress of the History of Science and Technology, which Berr also attended, as a stage for a performative display of the Marxist version of historical synthesis. Bukharin's performance is usually remembered as a watershed moment in the history of science, but it was also a turning point for larger debates about history, its method, its theory, and its relation to science.

This chapter traces the events leading to this moment of performative confrontation between some of the leading proponents of the refurbished versions of two major intellectual formations of the nineteenth century, positivism and Marxism, as manifested in their application to history and its method. By doing so, it introduces Russia as a site for the circulation, appropriation, and modification of the emerging knowledge, practices, and philosophies associated with scientific history. In the quest for scientific history discussed in the previous chapter, Russian historians were active, although unequal, participants. Their close ties to their Western European counterparts, combined with their preoccupation with comparative approaches within the broader context of debate about "Russia and the West," positioned them well in the international movement around scientific history.

Russia and the West

In Russia, professional history came of age in the second half of the nineteenth century under the sign of the great intellectual debate about Russia and the West, in which *West* stood, somewhat vaguely, for the West European model of governance and social organization.[2] At the heart of the debate was the question of Russia's relation with Western Europe and, more broadly, the nation's self-image vis-à-vis the West. Was Russia part of the West? Or was there a distinctive Russian culture built on its Orthodox religious tradition? The answers to these questions played an integral part in conceptions of Russian national identity. Since Peter the Great's reforms in the late seventeenth century, which formally changed the Tsardom of Muscovy into the Empire of all the Russias, the nation's intellectual elites were preoccupied with what became known as the great dispute between the so-called Slavophiles and the Westernizers about Russia's relation with the West and the meanings of its history.[3]

As members of intellectual elites who interpreted the meanings of broad cultural changes and specific political events, historians played a central role in this lasting debate about how Russia was to be defined. With the establishment of professional history in Germany, France, Italy, and Britain, Russian students of history became avid consumers of the new, scholarly discourse on the West European past. Specialists in Russian history studied historiographies written by their European counterparts to determine the place of Russia in the "world-universal histories" (*vseobshchaia istoriia*), which typically mentioned their nation only in passing, if at all. At the same time, there was a demand for Russia's own specialists in world history as Russian audiences clamored for access to lessons from the West European past. As Pavel Vinogradov, a Russian specialist on medieval England, put it at the close of the nineteenth century: "Everything in the West that is of interest to the antiquarian is of present-day value for us."[4]

The main universities in the Russian Empire had established university chairs for both Russian history and world history by the beginning of the nineteenth century. For the universalist historians, a prolonged trip abroad, usually for two years following completion of a university course in Russia, was a prerequisite for obtaining a teaching position as a professor at a Russian university. The practice was ended in 1848, for fear that Russian students would be exposed to the revolutionary ferment in Europe, but it was quickly restored when, in the wake of Russia's defeat in the 1856 Crimean War, Alexander II started a series of far-reaching reforms, culminating with the emancipation of the serfs in 1861, in an attempt to modernize his defeated country. Among the reforms was an 1863 decree instructing universities to send their graduates abroad to complete their education in the European centers of their specialization, with the costs covered by their universities or the Russian Ministry of Education.[5] Young Russian historians made their way across Western Europe, usually starting with Germany, their intellectual mecca, and then continuing through Britain, France, and Italy. As they returned to Russia and took up professorships of their own, they re-created the forms of academic life into which they had been initiated abroad, including the defining features of German historiographic scholarship—lecture, seminar, research paper, monograph, scholarly journal, graduate education, and two-step degree.[6]

The forms of academic life came along with the content. The work of Nikolai I. Kareev, the historian of the French Revolution who helped launch the nascent field of sociology in Russia is but one example. As other Russian historians did, Kareev traveled to France in 1877–78 to complete his education and conduct library and archival research for his dissertation on the peasantry on the eve of the French Revolution. Throughout his subsequent career, he regularly visited French, German, and other European universities, making connections with leading European historians, including Henri Berr, whose work he introduced to Russian audiences.[7]

As would so many other Russian intellectuals in the later decades of the nineteenth century, Kareev was an avid reader of Comte. After the revolutions of 1848–49, Comte's writings were removed from public libraries in Russia and effectively banned until the ascendance of the reform-minded Alexander II in the 1860s. After a period when the very words *society*, *revolution*, and *progress* were prohibited in official discourse, positivism, principally of its French variety, became widely influential in Russia.[8] Like other undisciplined followers of Comte, who reformulated and refashioned his positivist legacy in various ways, Kareev was neither a consistent nor a self-proclaimed positivist. He shared with other followers of positive philosophy a fundamental belief that a scientific study of society would bring about social change and an epistemic distaste for a neo-Kantian positioning of the natural and human sciences as a binary pair. As a theorist of history and the preeminent champion of the new science of sociology in Russia, he applauded Comte's division of disciplines and his notion of methodological unity.[9] As a historian, he believed in the virtues of comparison and applied the comparative method in his own work.[10]

At the close of the nineteenth century, professional historians in Russia shared a view that history was scientific only insofar as it was comparative. As the historian F. Zigel stated in 1899: "The historian needs to realize that one cannot simply describe without also juxtaposing, that history can be a science only inasmuch as it explains and there can be no explaining without comparing."[11] The *Journal of the Ministry of Education* (*Zhurnal Ministerstva narodnogo prosveshcheniia*), the premier Russian historical journal at the time, promoted comparative methodology and served as the main venue for comparative historiographies.[12]

Comparative history resonated so powerfully in Russia in part because it dovetailed well with the perennial "Russian question" of the nation's relation to the West. By uncovering parallels between the West's past and Russia's present, a generation of Russian historians who specialized in different aspects of European history and who came of age in the aftermath of Alexander II's reforms in the 1860s used comparative methodology to extract historical lessons for understanding the social and economic changes resulting from the Emancipation Act of 1861. Thus, for example, in his study of the French Revolution, Kareev focused on the "peasant question"—one of the most enduring questions in Russian thought. As he explained in his book on the peasantry on the eve of the French Revolution, he used the phrase *peasant question* in the title of his book to underscore his break with the traditional preoccupation of French historians with revolutionary elites and his shift of focus toward the living conditions of the French agricultural communities instead. "One would be searching in vain for the term *peasant question*, which I have chosen for the title of my book, in the vocabulary of French writers of the last century," Kareev wrote in the introduction. "This is because the peasant question, meaning the question of their [i.e., peasants'] existence as a class, their needs, interests, etc, had not yet taken root [in France]. By that I do not mean that the publicists and scholars simply ignored the peasants and their conditions. On the contrary, . . .

these issues were of utmost importance. But in their analyses they treated the peasants as part of broader social issues overlooking what constitutes the peasant question as such."[13]

Kareev's analysis of the French peasantry's worsening impoverishment and the general change in relations between the peasantry and the seigneurs had obvious relevance to audiences in a postemancipatory Russia interested in a scientifically grounded prognosis for future social and economic developments in their country. But it had found a receptive audience in France as well. Kareev's study of the peasantry on the eve of the French Revolution, which was translated into French during his lifetime, was well received by the historians of the *école méthodique* (see chapter 1 above).[14] Fustel de Coulanges, for instance, praised Kareev's perspective as an outside observer with a firsthand knowledge of serfdom, a position that allowed him to capture the nature of the relations between French peasants and their seigneurs in a way that his French colleagues could not. As Frances Nethercott has aptly put it discussing this episode, projecting a European past onto Russia's present was a "two-way mirror," "an invitation of sorts to the West to consider its own reflection in the Russian mirror."[15]

The salience of the Western European past for the Russian question in the present—"the challenge of modernity" (*vyzov sovremennosti*) in the common Russian parlance of the time—made Russian historians not only receptive to comparative methodology as a foundation of scientific history but also active participants in the international movement around scientific history detailed in the previous chapter.

Russian Historiography on the World Stage

The First International Historical Congress, which was organized during the 1900 Paris Exposition, turned out to be less international than intended, given that relatively few people outside France attended. Of about eight hundred registered participants, fewer than two hundred actually showed up. Few British and German historians were represented. Many of those who did attend came from Eastern Europe.[16] A handful of participants from Russia gave papers, but they otherwise stayed on the sidelines during the congress.[17] Back home, however, they reported on the new international organization for history at a meeting of the Russian Historical Society in Saint Petersburg.[18]

At the next international historical congress, which took place in Rome in 1903, Russia was represented by a large delegation headed by Vasilii Modestov, a prominent Russian authority on the history and culture of ancient Rome. Modestov was elected a vice president of the congress, alongside Adolph Harnack of Germany, Eduard Meyer of France, and James Bryce of England. Presiding over the congress was an honorary president in absentia, the venerable Theodor Mommsen, who died soon thereafter. Determined that Russian historians should send a dynamic delegation to the next meeting, scheduled for Berlin, one of the delegates at Rome, the medievalist Ivan Grevs, set out to organize a "preparatory congress of Russia's historians."[19] The preparatory congress never took place, but the Russian contingent of historians who came to Berlin in the summer of 1908 was among the largest of the foreign delegations, second only to those from England and Italy.[20] In Berlin, one of the delegates, the Russian-Ukrainian historian Nikolai Bubnov, insisted that Russian should, along with German, English, French, and Italian, be one of the official languages at the next congress, to be held in London in five years' time.[21] This was not to happen, although the 1913 congress in London did become a high point for Russia's historians on the world stage.

The member of the organizing committee who killed the proposal for Russian as an official congress language was, somewhat ironically, a Russian. A specialist in British medieval history, Pavel Vinogradov had elected to serve as a foreign member of the organizing committee for the 1913 international historical congress in London (the local organizing committees were traditionally made up of historians from the host country and one foreign member), having been nominated because, after accepting a professorship at Oxford, he moved to England while remaining affiliated with Moscow University.[22] In a letter discussing the issue of the Russian-language proposal addressed to his colleague in Saint Petersburg Aleksandr Lappo-Danilevskii, Vinogradov, writing from Oxford, vented: "This entire ploy is so characteristic and so ridiculous. Whom do these gentlemen [*gospoda*] think they attract by their recitals in a language [most of the participants] don't understand?"[23]

Addressing the session on legal history at the opening of the congress, Vinogradov stressed the virtues of comparative history. Recalling the first international meeting of historians, which had projected that comparative history would be the future of their discipline, he argued that only the "application of the comparative method" could elucidate the "streams of doctrines and institutional facts which pass through the ages and cross national boundaries from one historical formation to another." Comparative history, he continued, was the means with which to create "connecting links between the various researchers in Roman and English, German and Slavonic, Civil, Canon, and Common Law."[24]

The head of Russian delegation, Lappo-Danilevskii, who was in the audience, might have been nodding in agreement. He had established his reputation in Russia not only as a practicing historian who used the comparative method in his work but also as a theorist of historical method who devoted much effort to modernizing—that is, accounting for the deficiencies and contradictions in—the Comtean framework of positivism.[25]

Born in 1863, Lappo-Danilevskii grew up in an aristocratic family that encouraged his widely ranging interests in sciences and philosophy.[26] At Saint Petersburg Imperial University, he studied philosophy, law, physics, chemistry, astronomy, and mathematics alongside history, the social sciences, and economics. According to his friend Ivan Grevs, as a student of history he tried to work out a system of mathematical notation that could be applied to historical research, and for this reason he took all the mathematics courses the university offered.[27] He completed his first dissertation on the origin and material culture of the Scythians, having been drawn to the topic by the archaeological discoveries of the late nineteenth century that challenged historians' conventions about the boundaries of what qualified as history and historical method.[28] His interests later turned to the history of diplomacy, historical method, and the theory of history. In the latter area, he established himself as a recognized authority in Russia. Attracted to Comte and his followers by their bold quest for the unity of knowledge through unity of method, he engaged at the same time in an expansive critique of Comte and the earlier versions of positivism, warning against attempts to develop a science of history without a clear idea of up-to-date scientific methods.[29] In his monumental *Metodologiia istorii* (The methodology of history), first published in 1909, he presented his own synthetic exposition of scientific methods and ways in which these methods could be applied to the writing of history, making history a "rigorous science."[30]

In the *Metodologiia istorii*, Lappo-Danilevskii questioned the binary between natural scientific and historical methods, citing, in particular, Henri Poincaré's discussion of the limits of the natural laws in actual scientific practice (see chapter 1 above). He delineated several possible avenues for future scientific historians to take. One was a traditional historical analysis centered on individual historical events but sharpened by the use of various modern scientific techniques, such as rigorous linguistic analysis and the use of statistical analysis and computations. The introduction of such tools to their tool kit would, he argued, allow historians to widen their empirical base and replace "primitive observations" with mathematical "averages."[31] In his own work, Lappo-Danilevskii made a concerted effort to introduce nontextual sources, in particular material artifacts and such sources as drawings, paintings, sculpture, and architecture.[32]

Another avenue for scientific history that Lappo-Danilevskii outlined for historians was a "typological" approach, which drew on Comte's prescription that historical method was the last and most advanced part of the comparative method contributed by biology.[33] Like many other Russian intellectuals influenced by Comtean teaching, Lappo-Danilevskii saw biology as the paradigmatic science that contributed comparative methodology to a unified historical method.[34] However, he added a twist to the discussion, linking the comparative method to the sensibilities specific to the Russian locale.

Refurbishing Comtean ideas about comparison and history from a Russian perspective, Lappo-Danilevskii at the same time recast familiar themes in Russian historiography—the question of Russia and the West and the comparison between the European past and the Russian present. Applying the typological lens to the question of Russia's relation with the West, he argued that the great debate between the Slavophiles and the Westernizers emerged and was sustained in Russia as a manifestation of the rise of nationalism counterbalanced by the quest for internal unity and the integration of Russian society. Thus far, he argued, Russian society had been built on a borrowed culture and, therefore, had no internal unity. He held that the solution to the problem of Russia and the West was not to cast out Western culture in a search for genuine Russian national values but rather to integrate Western culture into a consistent system of Russian values. Accordingly, in his own work he shifted the focus from Western ideas in and of themselves to how these ideas were assimilated and "acclimatized" in Russia (a process he called *epigenesis*).[35] His methodological discussion aligned well with his political outlook as a Westernizer. A liberal and an internationalist, he saw international cooperation as one of the manifestations of a world order based on mutual appropriations, the order in which Russia, in its relation with the West, was but a typical example.[36]

The 1913 congress in London represented the pinnacle of international recognition for Lappo-Danilevskii personally as he was invited to deliver a plenary lecture. He was also elected the chairman of the organizing committee for the next congress, scheduled to take place in Russia's capital, Saint Petersburg, in 1918. As a general theme for the planned meeting, he proposed, fittingly, "the interconnection of Russia and the West."[37] That congress never materialized. World War I and the Russian revolutions of 1917 disrupted the planning. The next inter national congress did not meet until 1923. By then, Russia was a different country. On December 30, 1922, a treaty signed between Russia, Ukraine, Belarus, and Transcaucasia (a region between the Black and Caspian Seas that constituted a short-lived republic) formed the Union of Soviet Socialist Republics.

Historians in Russia were affected by the revolutions of 1917 both as citizens and as a professional group. Large numbers of them actively participated in the revolutionary movement. During the revolution of 1905, the historian Pavel Milyukov formed the Constitutional Democratic Party or Kadet Party, an organization usually described as a stronghold of Russian liberalism, which meant, at the time, socialist commitments with a strong preference for reform over revolution. The Kadet Party distinguished itself from other political parties by emphasizing gradual change through reform and education, rejecting violence as a means of political struggle.[38] It was, to a large degree, the party of historians. Kareev, Grevs, and Lappo-Danilevskii as well as most of the other leading Russian historians were Kadets. Most of them supported the February Revolution of 1917, and many served in the short-lived Provisional Government, which replaced the tsarist government only to be overthrown by the Bolsheviks in October of the same year. That political engagement left them all especially vulnerable to retaliation after the coup. The situation was exacerbated by the fact that one of the leading Bolshevik reformers of the humanities was a professional historian, Mikhail Pokrovskii.

When international contacts resumed after a decade-long interruption, the Soviet historians-cum-politicians started to claim their place on the world stage through the venue of international historical congresses, showcasing Marxist history as a new brand of scientific history. Pokrovskii was its first spokesman.

Marxism and History

As soon as the Bolsheviks emerged victorious from the merciless civil war that ravaged the country for more than four years, the new leaders identified history as a new "battleground"—the "third," "cultural," front of the Revolution.[39] As a locus of the revolutionary struggle, the cultural revolution aimed to transform people's consciousness by building a new culture, creating a new science, shaping the revolutionary scholars, and molding "new people." History was an important part of this reconfiguration of consciousness.

The close-knit circle of the revolutionary vanguard that formed the core of the Bolshevik wing of the Russian Social-Democratic Labor

Party had laid the groundwork for the new type of scholarship for more than a decade before seizing power.[40] In the wake of the suppression that followed the first Russian revolution of 1905, Vladimir Lenin and other leaders of the Bolshevik faction established underground party schools in Capri, Bologna, and Longjumeau, a one-street village outside Paris where Inessa Armand, a feminist and member of the Bolshevik Party (and Lenin's mistress) rented a metalworkers' shop for the underground party university. For a while in the early 1910s, these party schools functioned as a political academy for Russian workers, offering training in the organization of propaganda and agitation for the revolution. The education process itself was revolutionary: the schools were run by a committee composed of students and teachers who cooperatively worked out the curriculum. In both their form and their content, the schools were the testing ground for a new system of education.[41]

Among the lecturers who taught in all three party schools was Mikhail Pokrovskii, the only professional historian among the leading Bolsheviks. Throughout the 1920s, and until his death (from cancer) in 1932, Pokrovskii was the most influential Marxist historian in the Soviet Union. He was the first to write a complete course of Russian history from a Marxist perspective, and he established institutes for the new Marxist historiography, ultimately training what Vladimir Bonch-Bruevich, Lenin's personal secretary, described as a "galaxy of new Marxist historians."[42]

The official dean of the Soviet historical profession in the making, Mikhail Pokrovskii was born in 1868 into the family of a customs official. After graduating from a classical gymnasium in Moscow, he studied history at the Historical-Philological Faculty of Moscow University. He attended seminars of renowned Russian historians, including those of Vinogradov and Vasilii Kliuchevskii. After completing his master's (magister) degree in 1894, he took a teaching position in a pedagogical institute. He published his first scholarly work in a series on medieval history edited by Vinogradov, who solicited the contribution.[43] Having proved himself a talented scholar, Pokrovskii was offered the prestigious opportunity to "prepare for a professorship," which in pre-Revolution Russia meant an appointment at a university granted only to candidates recommended by outstanding scholars in their fields. He declined the offer.[44]

By the time he received this offer, Pokrovskii was actively engaged in the revolutionary movement. He joined the Russian Social-Democratic Labor Party on the eve of first Russian revolution of 1905 and soon became part of Lenin's circle, within which he was nicknamed *the profes-*

sor with a lance, a label that stuck with him until the end of his life. He was a prolific contributor to the Bolshevik newspapers *Pravda* and *Rabochii*. Among his contributions to *Pravda* was a review of Kliuchevskii's *Course of Russian History* in which he criticized his former teacher for the "eclecticism" of his theoretical views and for neglecting Marx's theory of history. Chased by the police, he fled the country in 1907, moving first to Finland and then to France. He lived in exile until 1917.

Exile turned out to be a time of intense scholarly activity for Pokrovskii. As a lecturer at the party schools, he taught Russian history to future "agents of revolution," turning from critiquing his university history professors to writing a radically new historiography. His aim, as he explained in the introduction to his five-volume *Russian History from the Earliest Times* (1910–13), was "to reinterpret Russian history from the Marxist point of view," reusing and reinterpreting material assembled by "bourgeois historians."[45] He went on to produce a corpus of work offering a radical reinterpretation of Russian history: *Sketches in the History of Russian Culture* (1914–18), *Brief History of Russia* (1923), and *Sketches in the History of the Russian Revolutionary Movement* (1924).[46]

Criticizing historians for what he saw as flawed analyses of the historical record through the lens of nation and national identity, Pokrovskii called on a new generation of Marxist historians to reexamine the historical record through the lens of class and capital. Only such a Marxist, class-based, analysis could be truly scientific, he argued. History, he stated in an often-quoted line, was "the most political science of all," but history "written by these gentlemen [i.e., bourgeois historians] is nothing but politics projected into the past."[47]

Taking a stab at the familiar subject of Russia and the West, Pokrovskii forcefully argued against the view that Russia had a unique historical path. He made his case by analyzing Russian history since "the gathering of the Russian lands" in medieval Muscovy through the lens of trade and "merchant capital" as the drivers of the domestic and foreign policies of the Romanovs. What his Marxist, class-based analysis unmasked, he argued, was that Russian empire was built on exploitation and looting, driven by the promise of new markets and new trade routes. After abolishing serfdom, Alexander II conquered new territories primarily to open new markets. Hence, he chose a region that "seemed far away from all rivals of Russia—Central Asia, whose immediate neighbors were the Chinese and the Afghans": "The conquest of Central Asia had a huge significance for the development of Russian industry. It became the first Russian colony."[48] As a result: "The Romanov's Empire . . .

united under a single power the most diverse people who had the misfortune to live near the trade routes needed for Russian merchant capital."[49] The huge territory that was conquered, which stretched from Moscow to the Pacific Ocean, "was called 'Russia,'" and all the people therein were "ordered to call themselves 'Russian,'" despite the fact that most of them "did not understand a word of the language spoken in Moscow."[50]

Until the end of his life, Pokrovskii vehemently denounced Russia's imperial past. Soon after his death, however, his lambasting of "Great Russian chauvinism" started to sound increasingly out of sync with the newly adopted Russocentric "national policy" (*natsional'naia politika*) introduced by Stalin in the early 1930s in the aftermath of his brutal collectivization campaign. In January 1936, four years after Pokrovskii's death, the Party's Central Committee denounced the "false historical views professed by the so-called historical school of Pokrovskii" in a resolution published in leading Soviet newspapers.[51] During his lifetime, however, not only was Pokrovskii revered as the Soviet Union's leading Marxist historian, but he also enjoyed an exceptional status in the Soviet hierarchy of power arguably unmatched by any historian before or after.

In the 1920s, Pokrovskii designed and launched reforms aimed at revolutionizing the teaching of history and training a new generation of scholars-cum-revolutionaries. In early 1918, the People's Commissariat of Enlightenment, or Narkompros, began to reform the new socialist state's universities and scientific institutions. Early Soviet policy toward existing educational and scientific institutions differed for the natural sciences and for the humanities. For the former, the policy adopted sought to co-opt existing scientific-technical expertise.[52] The social sciences and humanities, on the other hand, were to be replaced with the scholarly production of the new Communist institutes, created to train a new generation of Communist intelligentsia. In 1922, Lenin ordered the deportation of 160 Russian scholars, including some of the best-known and influential historians.[53] As Narkompros's vice chair, Pokrovskii was in charge of reforming education in the humanities and training replacements for the "bourgeois academics." He established or helped establish a host of new institutes, including the Institute of Red Professors, the Communist Academy, Lenin's Institute (created after Lenin's death in 1924), the Marx-Engels Institute, and the Communist University (or Sverdlov University).[54]

All these new centers offered instruction in history. The Institute of Red Professors alone admitted 157 history students between 1921 and 1928, effectively creating the first generation of Soviet Marxist historians.[55] In 1925, to provide a scholarly center for Marxist historiography, Pokrovskii established the Society of Marxist Historians. He also created a host of new journals, including the influential *Istorik-Marksist* (Marxist historian), devoted to the theory of history and questions of method, the *Krasnyi arkhiv* (Red archive), a venue for publishing historical sources, and *Klassovaia bor'ba* (Class struggle), an outlet for historical writings focused on revolutionary subjects such as the history of socialism, the history of the workers' movement, and the history of the Revolution itself.[56] In March 1928, at the First Conference of Marxist-Leninist Scientific Research Institutions, Pokrovskii declared that a school of Marxist historians had been created.[57] In the summer of that year, he led a delegation of his trainees to the international historical congress in Oslo.

After the 1913 London meeting, the congress did not meet again for a decade. When it finally did meet in 1923 in Brussels, the history profession in Russia was represented at the meeting by the same people as before, only in lesser numbers.[58] Three of them (Mikhail Rostovtsev, Petr Struve, and Vinogradov) were political refugees who had fled Russia after the Revolution. Three other Russian historians came from Petrograd (as Saint Petersburg had been renamed at the outbreak of World War I, only to be renamed Leningrad in 1924). The mixed group registered as representing the Russian Academy of Sciences rather than the Soviet Union, an entity not yet recognized by most European governments. Moreover, Vinogradov and Rostovtsev continued to represent Russia on the organizing committee, a situation that the committee's chairman, the American historian Waldo Leland, found absurd, making it a point to have Soviet scholars at the next meeting, scheduled for Oslo in 1928. In a private letter, he wondered how Russian historiography could possibly be represented by one scholar who had been based in Oxford since 1903 (Pavel Vinogradov, by then better known as Paul Vinogradoff) and one who had been based in the United States since 1920 (Rostovtsev): "We must deal directly with Russian scholars in Russia and not adopt the fiction that they can be represented by Russian scholars outside Russia."[59]

The organizers of the Oslo congress formally invited the Academy of Sciences in Leningrad to send representatives to the planning meeting in Geneva. They received a response: the Soviet historians could not come to Geneva since Switzerland did not recognize the Soviet Union, but a delegation would come to Oslo.[60] The invitation was answered by none other than Pokrovskii, who was organizing the "week of Soviet historical science" to be held in Berlin in July 1928, just before the historical congress scheduled to meet in Oslo in August.[61] The timing of the showcasing of Pokrovskii's school of Marxist history abroad was, indeed, fortunate. After years of shattered international contacts, with both Germany and Russia politically isolated by the terms of the Treaty of Versailles, the Locarno Treaties of 1926 had brought Germany back into the community of nations. The Bolshevik leaders rightly saw an opportunity for the Soviet Union as well and actively sought to restore international contacts, using scientific internationalism as a means of international diplomacy.[62] Pokrovskii lumped plans for Berlin and Oslo together into a high-profile tour featuring his school's work outside the Soviet Union for the first time. Besides Pokrovskii and the first graduates of the Institute of Red Professors, Issak Mints and Sergei Dubrovskii, the delegation included several high-profile Communist politicians.[63] For many Westerners, that congress would be their first encounter with Communist scholarship.

At the congress in Oslo, reconciliation was the order of the day. For the first time since World War I, the international gathering of historians featured groups from countries that not long ago had fought against each other on the battlefield. Pokrovskii and his team were greeted with accentuated courtesy by the congress organizers, who urged the delegates, in the words of the chairman of the organizing committee, Halvdan Koht, to endorse "the will of mutual understanding" as "the spirit of Oslo."[64] The Koht-led committee proceeded to elect Pokrovskii as one of the congress presidents. He was also invited to deliver a prestigious plenary lecture at the closing session. He used the opportunity to take on the well-worn subject of Russian absolutism, usually treated as a manifestation of Russia's somewhat unique historical path, and present a Marxist perspective on the issue. Answering his own question—Was Russian absolutism different in kind from the European styles of governance?—he argued that the Russian form of governance was not different in kind from that of other imperialist states and that empires are best understood as agents of trade capital in the struggle for markets in Eastern Europe and Asia.[65]

The Oslo congress set up important contacts between Soviet Marxist historians and their Western counterparts. For Pokrovskii personally, the congress was a triumph. He took no joy, however, from his disciples' performance at the meeting and used the experience to establish a rule of conduct for Soviet delegations in years to come. After returning to Moscow, he castigated the members of the delegation for their passivity. In the report on the congress published in the bulletin of the Communist Academy, he wrote: "We come from the country that was the first to introduce the dictatorship of the proletariat, we come, sit, and remain silent. That was our big mistake, the biggest of all." Regarding those who simply presented their own papers and did not participate in the discussions, he lamented: "Everyone must make an appearance in the panels of their area of history, and make two or three critical contributions, and prepare for them. I stood up twice, not particularly aggressively, but others remained entirely silent." This passive style had to change, he argued. Marxist historians must perform as scholar-revolutionaries on both the domestic and the international fronts. On the basis of the delegation's performance in Oslo, he proposed guidelines for how Marxist historians should perform at future international gatherings. First, they must intervene in all panels, which would require "outstanding language abilities"; second, they must "achieve agreement and unity of opinion within our own ranks"; and, third, they must challenge and "unmask" the "aura of objectivity" assumed by Western historians, which would require an "excellent mastery of the theory and methodology of bourgeois historical scholarship."[66]

Pokrovskii's experiences at Oslo may have been colored by his personal circumstances. In the preceding year, 1927, he had been diagnosed with the bladder cancer of which he would die in April 1932. By the time of the Oslo congress, he was already seriously ill and under medical treatment. After Oslo, he returned to Berlin (where he had just been for the week of Soviet historical science). This time, he checked into a clinic and had surgery. On his return to Moscow, he wrote a piece for a leading Soviet newspaper, describing the fight for history in near-apocalyptic terms: "The fight is going on on the international front, and our obligation as Soviet historians is to use all our resources to defend our stance on a global stage. . . . The historical front is part of the ideological front, which in itself is part of the general preparation for the uncontrollable cataclysm that is heading toward us."[67] Whatever internal state might have prompted Pokrovskii's high-pitched tone, the characterization of a "fight" that pitted Marxist historians against their bourgeois class enemies reflected Soviet officials' views on the stakes of history in the dramatic period in Soviet history that Stalin dubbed the "Great Break."

The Great Break

On November 7, 1929, one year into the first Five-Year Plan, *Pravda*, the mouthpiece of the Party and the leading Soviet newspaper, published an article by Stalin titled "The Year of the Great Break." In it, Stalin heralded the Five-Year Plan as the beginning of the last great revolutionary struggle against "capitalist elements" within the Soviet Union. What he meant by the "Great Break," as he explained to Maxim Gorky in a letter, was the "total breaking up of the old society and the feverish building of the new."[68] The struggle was to encompass all spheres of life, science included.

The rhetoric of "two sciences" mirroring "two ways of life," new and old, "theirs" and "ours," had been in circulation as early as 1918. A resolution on the "reforms of scientific institutions and schools of higher learning" issued by the union of the communes of the Archangel region in the first post-Revolution year, for instance, called, extravagantly, for an outright liquidation of the old institutions—the "unusable relics of the pseudoclassical epoch in the development of class society." As the resolution stated: "Communist science can be thought of only as a collective and vital deed of all people and their labor." In contrast to this ideal, the old scientific institutions were but "the shamanism of the magi in their exclusive sanctuaries, the sinecures, which can cultivate only the class psychology of virtuous priests or charlatans."[69] The first Communist textbook, *The ABC of Communism*, written in 1919 by Nikolai Bukharin, presented the newly established Communist Academy as a revolutionary alternative to universities and the old, imperial Russian Academy of Sciences. It was to be "a new type of school, serviceable to the revolution now in progress."[70] With the launch of the first Five-Year Plan, the rhetoric of two sciences intensified. The Great Break signaled that the revolutionary forces of the new world had entered the "final struggle" with the relics of the old one.[71]

As the head of the new Communist institutions, Pokrovskii immediately translated the new terms for any representatives of the so-called bourgeois science left in the country: "The period of peaceful coexistence with bourgeois science is over. It is time to go on the offensive on all scientific fronts."[72] Pokrovskii had claimed that he was calling only for symbolic violence, not physical terror ("I hope that nobody will understand [the opening of] fire literally; we are positively not going to inflict any physical damage on anyone").[73] But violent actions quickly followed the violent rhetoric. A wide array of groups in all academic, cultural, and technical professions found themselves under attack. The Shakhty trial (*Shakhtinskoe delo*) of engineers in the spring of 1928 opened a series of show trials that targeted non-Communist specialists on the (false) grounds that they had formed clandestine anti-Soviet organizations and parties plotting a coup against the Soviet government.[74] The first case featured prominent historians.

The case, which involved fabricated accusations against the Academy of Sciences at large (*akademicheskoe delo*, "the Academic affair"), never came to trial, but hundreds of professors were arrested and lost their jobs, and many had to leave Leningrad. The pretext was the academy's record in the spring 1928 election of new members, when only five of eight Communists nominated by the Politburo gained seats. Those elected included the prominent Bolshevik leader and head of the Comintern Nikolai Bukharin; the Red historians Mikhail Pokrovskii and the founder of the Marx-Engels Institute, Dmitrii Riazanov, and two engineers among the Bolshevik elite, Gleb Krzhizhanovsky and Igor Gubkin. Three other Communist candidates, including two historians from Pokrovskii's delegation to Oslo, Nikolai Lukin and Vladimir Friche, as well as the Communist philosopher Abram Deborin, were turned down. When the depth of the error became clear, the academy held a special ad hoc election. In February 1929, the remaining candidates were voted in. In return, the academy's budget was almost doubled.[75] Yet the stage was set for an instructive purge. Pokrovskii himself brought the matter to the Politburo's attention.[76] In what followed, historians featured prominently: theirs was the only professional group implicated in the Academic affair who actually went on trial.

It all started with an archival find. In January 1929, during a routine inspection by the newly organized Central Archive of the material held in different libraries under the jurisdiction of the Academy of Sciences in Leningrad, a treasure trove of documents that chronicled the Revolution of 1917 preceding the Bolshevik takeover was found in the library of the Pushkin House. Included were the original records of Nicholas II's abdication, material relating to the short-lived Provisional Government, the papers of Petr Struve and other leaders of the Kadet Party, documents relating to the Central Committee of the Social-Revolutionary Party, and so forth. The material came from different sources, often delivered to Pushkin House during the turmoil of the Revolution and the Civil War by owners who expected or demanded discretion. Some of the documents, such as the personal papers of the head of the tsarist secret police, Alexander von Benkendorff, were handed over in sealed boxes that were never opened by the library staff under the terms of an agreement according to which the contents were not be disclosed for twenty years. The contents of those boxes, as well as other material kept at the Pushkin House, were revealed by the Central Archive inspectors. The controversy was exaggerated by the ongoing rivalry between the academy's archivists and the newly formed Central Archive, headed by none other than Pokrovskii. What started as a quarrel between the academy's historians and the staff of the Central Archive about the proper way to preserve historical documents quickly grew into political accusations in the context of the recent elections and the short-lived display of resistance by the academy. The OGPU (the Soviet political police) set up an investigating commission that started issuing arrest orders.[77]

By the end of 1929, as many as 115 leading Russian historians from Leningrad, Moscow, and other cities were arrested in relation to the Academic affair.[78] Among these was the academy's "permanent secretary" Sergei Oldenburg, the prominent historian-Orientalist. One of the leaders of the Kadet Party, Oldenburg had served as the minister of education in the Provisional Government after the February Revolution and then negotiated a deal with the Bolsheviks to keep the Academy of Sciences autonomous in return for the services the "bourgeois" specialists provided to the new regime. Now he was accused of serious negligence and sent into retirement in disgrace at age sixty-eight. Those who were arrested stood trial. The accusers used their pre-Revolution activities as members of the Kadet Party to fabricate a conspiracy involving a mythical counterrevolutionary organization called the "People's Union." Most of those arrested were ousted from their positions, received prison sentences, or were sent into exile.[79]

As professional historians were removed from their positions or otherwise silenced, politicians took their place as interpreters of the historical meaning of contemporary events.[80] The leading Bolshevik theoretician Nikolai Bukharin not only offered historical interpretations but also engaged with the academic debates on history, its method, its theory, and its relation to science. Bukharin's foray into academic debates was prompted by his own downfall as a politician in the aftermath of the Great Break. In April 1929, he was ousted from his high-ranking positions as the chairman of the Comintern and the editor of *Pravda*; in November, he was removed from the Politburo, the highest authority within the Communist Party. He was assigned to head the Scientific-Technical Division (Nauchno-tekhnicheskii otdel, or NTO) of the Supreme Council of the National Economy (Vysshii sovet narodnogo khoziaistva, better known by its acronym, VSNKh), a position from which he began debating how, why, and what to plan in science and technology. From 1929 until 1934, when he was ousted from his position at the NTO as well, he tried to influence decisions on policies and the implementation of the first and second Five-Year Plans, only to fall out with Stalin, which contributed to his ultimate downfall.[81]

It was during the years between 1930 and 1933, when his political role was his least important since the Revolution, that Bukharin became actively involved in the history of science. In October 1930, amid Stalin's campaign against him, he was assigned to oversee the Commission on the History of Knowledge (Komissiia po istorii znaniia, or KIZ), another appointment that marked his political downfall.[82] He found himself at the head of an institution devoted to the history of science broadly conceived that had begun to establish connections with international organizations. Two members of commission—the historian of ancient Egypt Vasilii Struve and the chemist Max Bloch—served as corresponding members of the International Committee of the History of Science, which was affiliated with Henri Berr's International Center of Synthesis, an organization dedicated to the unification of knowledge (see chapter 1 above).[83] As the new director of the KIZ, Bukharin would eventually meet the dean of historical synthesis at the congress of science historians in London, which Struve planned to attend in the summer of 1931.

Bukharin and the History of Science

The KIZ was founded in 1921 under the auspices of the Russian Academy of Sciences, Russia's oldest scientific institution.[84] The driving force behind the KIZ was the noted geochemist Vladimir Vernadskii. Not unlike Henri Berr, Vernadskii saw the history of science within the larger framework of the unity of knowledge, a safeguard against specialization and a natural bridge between the sciences, philosophy, and the humanities.[85] He invested the history of science with additional meanings, however, replete with theosophical and cosmist ideas couched in the rhetoric of more conventional intellectual discourse.[86] The immediate rationale for the founding of the KIZ in the midst of the Civil War and post-Revolution chaos was also perhaps pragmatic, driven by Vernadskii's concerns over the survival of the pre-Revolution intelligentsia and their values. In the immediate aftermath of the Revolution, many militant Bolsheviks called for the Academy of Sciences to be abolished as a remnant and the embodiment of Russia's "bourgeois, imperialist past."[87] By celebrating the history and achievements of Russian science since the formation of the academy by Peter the Great in 1725, Vernadskii hoped to demonstrate the value of the nation's internationally renowned scientific institutions to the new regime, strengthening the case not only for the preservation of the academy but also for its autonomy.[88]

Soon after founding the KIZ, however, Vernadskii, who reasonably feared for his life as a former member of the Provisional Government and a prominent Kadet, left Russia and moved to Paris, where his son George had already settled, having fled Russia soon after the Revolution.[89]

He had intended to remain permanently in France, but his attempts to obtain the funds to establish a laboratory of his own failed. In 1926, aged sixty-three, he returned to the Soviet Union. A highly regarded scientist and an institution builder who created and headed organizations critically important for the Five-Year Plan, including the Commission for the Study of the Natural Productive Forces of Russia (Komissiia po izucheniiu estestvennykh proizvoditel'nykh sil Rossii, or KEPS, established in 1915), he quickly assumed a role as one of the leading mediators between academics and the Bolshevik government.[90] As a technical specialist important to the regime, he was not persecuted during Stalin's Great Break. While his close friends, such as the historian Oldenburg, were harassed and arrested during the fabricated Academic affair, Vernadskii kept most of his scientific positions. The KIZ, however, as an organization that included a number of historians, including Oldenburg, was another matter.

In the early spring of 1930, Vernadskii learned, to his dismay, that Bukharin had been appointed to replace him as the director of the KIZ. The members of the commission who represented Russia on the International Committee of the History of Science found themselves in a particularly precarious situation. Vasilii Struve, who had just published (in German) an extended study of the so-called Moscow Mathematical Papyrus, was scheduled to attend the forthcoming Second International Congress of the History of Science and Technology.[91] Under the circumstances, he promptly canceled his plans.[92] The nephew of the prominent Kadet and now political emigré Petr Struve, Struve undoubtedly felt threatened. In the following years, he abruptly shifted his research agenda and relaunched himself as a Marxist.[93]

Bukharin, in the meantime, went to great lengths to cultivate collaborative and friendly relations with his predecessor and win the confidence of the KIZ members.[94] In February 1931, as the head of the KIZ, he received the circular letter about the forthcoming congress of science historians in London.[95] He presented the matter at a KIZ meeting and requested the commission's secretary to find out more about the congress.[96]

Simultaneously, Bukharin began to lobby the Politburo to send a large, high-profile delegation to London. In April, writing to the chairman of the Soviet of People's Commissars (or Sovnarkom, the Soviet government), Vyacheslav Molotov, he requested that the question be put "urgently" before the Politburo, stressing that the congress "is being organized very broadly (the Americans, French, Germans and Britons will be [represented in great numbers], with America sending a very large delegation)" and that "the themes are very interesting for us." He singled out two of the congress's themes: "the sciences as an integral part of general historical studies (very interesting for us in relation to Marxist newly published work [i.e., Friedrich Engels's *Dialectics of Nature*, a collections of unfinished writings on science that was first published by the Marx-Engels Institute in 1925])" and "the interdependence of pure and applied science (this last theme is utterly attractive for us)." He stressed: "For political, as well as all sorts of other reasons, it is necessary to send a substantial delegation."[97] The Politburo concurred. The delegation, approved in early June, included two engineers closely involved with the Five-Year Plan industrialization effort (Vladimir Mitkevich and Modest Rubinstein), Marxist scholars representing the Communist Academy (the philosopher-mathematician Ernest Kolman, the philosopher-physicist Boris Hessen, and the philosopher-biologist Boris Zavadovskii), and two non-Marxist scientists representing the Academy of Sciences (the physicist Abram Iofee and the biologist Nikolai Vavilov).[98]

The timing of the event might explain Bukharin's sudden interest in an international gathering of historians of science.[99] By 1931, Stalin had become anxious about the chaotic progress of the Five-Year Plan. In June 1931, he made a public announcement in which he stressed the need to attract "bourgeois specialists" to help in the industrialization effort. The London congress presented an opportunity to establish contacts with specialists in different areas of science and technology, an opportunity that Bukharin duly exploited. In the report he submitted to Stalin after his return from London in late July 1931, he wrote: "Perhaps much more important than the Congress itself was making contacts with English scientists."[100] During the congress Bukharin also met with Russian scientists who had left Russia after the Revolution and pursued successful careers in England.[101]

Besides immediate political goals, however, Bukharin also had an ambitious intellectual agenda. In the 1920s, he established his reputation as the Party's major theorist and the foremost Soviet systematizer of Marxism. In his major theoretical work, *Historical Materialism*, published in 1921,[102] he presented a synthetic account linking Marx's ideas to contemporary developments in the social and natural sciences and discussed a major challenge to Marxism, one coming from sociology. From its beginning in the Comtean system of positive philosophy, sociology had been developed into an exciting new science by Émile Durkheim, Vilfredo Pareto, Benedetto Croce, Max Weber, and other major Western figures and was being closely followed in Russia. Instead of dismissing "bourgeois sociology," as many other Marxists did at the time, he characterized it as "very interesting" and went on to reconcile Marx's system with the best of Comte's. Echoing Comte, he wrote: "Among the social sciences there are two important branches which consider not a single field of social life, but the entire social life in all its fullness. . . . One of these sciences is history; the other is sociology." While "history investigates and describes how the current of social life flowed at a certain time and in a certain place," sociology "takes up the answer to general questions." As "the most general (abstract) of the social sciences," it "formulates a method for history." In this light, Marxian "historical materialism," Bukharin stated, "is not political economy, nor is it history; it is the general theory of society . . . , i.e., sociology."[103] The title of the book encapsulated his thesis succinctly: *Historical Materialism* was subtitled *A Popular Textbook of Marxist Sociology*.

The Comtean overtones of his exposition as well as the very attempt to reconcile Marx with bourgeois sociology did not play out well for Bukharin. With many similarities between Marxism and positivism, from the law of the stages of development to the notion that the political system reflected the systems of knowledge, Marxists have been keen to differentiate themselves from their great intellectual rival, beginning with Marx himself and his disparaging remarks about "this shit-Positivism."[104] In 1930, during the anti-Bukharin campaign, the designation of historical materialism as sociology became an official example of political deviation.[105] As one critic wrote in a piece attacking Bukharin: "The representation of Marx as an advocate of a 'sociological method' can only lead to a rapprochement of his teaching with the teaching of bourgeois 'sociologists,' which has nothing in common with Marxism."[106] In November 1930, after the central press officially decried "Bukharin's conspiracy of silence" in response to such accusations, Bukharin signed a statement in which he acknowledged his "mistakes" and "deviations from the party line."[107] While still hoping for a political comeback, he made a major speech in London on theoretical issues of Marxism while plainly differentiating himself from Western bourgeois scholarship.

London 1931

As soon as the Bukharin-led delegation landed at Croydon Aerodrome near London on June 25—a rare transcontinental flight that itself drew attention to this event—the arrival of the Bolshevik politician in the British capital made political headlines. Right-wing newspapers used the visit as a weapon to implicate Ramsay MacDonald's Labour government, which had issued entry visas to the Soviets and was accused of helping disseminate Soviet propaganda and making concessions to Bolshevism, that "ominous shadow overhanging the peace of Europe." The *Daily Mail* screamed: "Bukharin, Head of Hate Factory!" "Moscow Hater of Britain now in London!" the *Morning Post* seconded. "One of the most virulent of communist agitators" had been admitted to the country as a "professor"![108]

And Bukharin did not disappoint. In the paper (delivered in German) that he presented at the congress, he spoke of "two worlds," "two cultures," irreconcilable in their ideological, economic, and existential opposition to each other. As he put it in the English version of the paper: "The whole of humanity, the whole *orbis terrarum*, has fallen apart into two worlds, two economic and cultural-historical systems. A great world-historic antithesis has arisen: there has taken place before our eyes the polarization of economic systems, the polarization of classes, . . . the polarization of cultures."[109]

In his London paper, Bukharin repeated the points he had made in a speech delivered in Moscow before leaving for London. On June 21–27, the Communist membership of the Academy of Sciences had held a special meeting to discuss what science could contribute to the realization of the motto "To Catch Up and to Overpass the Capitalistic Countries?"[110] Bukharin's speech was unambiguously titled "The Battle between Two Worlds and the Goals of Science: Science in the Soviet Union at the Great World-Historical Crossroads." As he did in his London paper, he began by declaring the unprecedented novelty of the present moment in history:

> All previous breaks and zigzags along the historical paths, all known historical crossroads—be they the great peasant revolutions of ancient China or the fall of [Greco-Roman] antiquity, the collapse of ancient Eastern despotism or the bourgeois revolutions, starting with the English one, the epoch of the Renaissance and the so-called Reformation or the dissolution of Genghis Khan's kingdom, the largest of all kingdoms—none of these milestones of the world-historical itinerary of humankind can be compared, neither qualitatively nor quantitatively . . . , to the historical shifts that characterize the time we live in now.

The conclusion of the speech reiterated the slogans of the Great Break: "The world has entered the phase of the last and decisive battles. The world is being split. There are two worlds now. Science is being split.

There are two sciences now. Culture is being split. There are two cultures now."[111]

In Moscow and London, Bukharin sandwiched a more sophisticated message between the slogans: that history is the key to everything and that science is embodied within society, inseparable from it. One consequence of this Marxist mode of thinking, he stressed, was that the opposition between "physical and intellectual labor," between "theory and practice," between "the specializations fragmenting and separating different sciences," and between "all of science from all of practice," vanished.[112] In Moscow, he underscored that successful planning requires an understanding of this hitherto obscured but very material interconnectedness of science with the social world. In London, he offered a series of arguments intended to convey that the Soviet regime had a deep appreciation of the unity of the sciences and recognized the interconnectedness of different fields of knowledge, of science both pure and applied, of science and technology.

Bukharin, it seems, understood very well the significance of the London congress in the context of the intellectual movement for the unification of knowledge and tried to play both sides. In London, he presented a sophisticated discussion weaving in quotations from an array of philosophers from Francis Bacon and George Berkeley to William James, Henri Bergson, and Max Scheler. All these thinkers, he admitted, sought a unified system of knowledge. Yet, peppering his text with emphases, he argued: "The *Socialist* unification of theory and practice is the most *radical* unification. . . . We are arriving *not only at a synthesis of science* but at *a social synthesis of science and practice*."[113]

It is hard to say whether Bukharin had Berr, the dean of historical synthesis, in mind when he was writing his paper. Indirect evidence suggests that he and Berr met in London or at the very least communicated through mediators. Berr attended the congress, participating, as a "member of honor," in the meetings of the International Committee of the History of Science held concurrently with the congress, much as such meetings had been held during the first congress in Paris in 1929.[114] One of the items on the commission's agenda was the planning of a session on the history of science for the next international historical congress, to be held in 1933 in Warsaw.[115] Soon afterward, both Berr and Bukharin submitted their proposals for Warsaw, where they were to appear on the same panel, "Historical Methods and Theory of History."[116] The preliminary arrangements were most likely made in London. Bukharin, who was in the midst of a political crisis at home, withdrew from the congress at the last moment.[117] With his name still listed in the program, the panel drew an unusual crowd of participants and outside observers curious to see a Soviet politician cross rhetorical swords with the bourgeois historian over alternative visions of historical synthesis.[118] The actual panel that took place at the congress was a mere shadow of the envisioned great spectacle of performative confrontation.[119]

Yet Bukharin still managed to stage a one-sided rhetorical fight with Berr in a publication he put together that year, *Marxism and Modern Thought*. The collection of essays, published in English in 1933, originated in a meeting between Bukharin and the British journalist James Crowther, a frequent visitor to the Soviet Union. During a visit in early 1933, Crowther asked Bukharin for suggestions of "recent works which, if translated, would help provide readers of English with a better understanding of the intellectual basis of the reconstruction of society progressing in the USSR."[120] In response, Bukharin proposed publishing a selection of essays written by leading Soviet academics for the occasion of the fiftieth anniversary of the death of Karl Marx. The authors he selected included the physicist Sergei Vavilov, the biologists Vladimir Komarov and Yakov Uranovskii, the Marxist philosopher Abram Deborin, and the historian Aleksandr Tumenev.

Bukharin's own essay and that of Tumenev, which bookended the collection, are of particular interest. In his essay, Bukharin engaged in a heated polemic with bourgeois philosophers, contrasting Marxism to all major philosophical trends of the time: logical positivism, pragmatism, Gestalt psychology, and so forth.[121] Tumenev's essay emulated the structure of Bukharin's, focusing on historiographic trends. A disproportionately long section of it is devoted to a critique of Henri Berr and his program of historical synthesis, focusing, in particular, on Berr's International Center of Synthesis, whose origins Tumenev dated, erroneously, to 1900. Other of Berr's projects, including the center's plans for the publication of "a dictionary of historical terms," were all listed. In his critique, Tumenev zoomed in on the multivolume *L'évolution de l'humanité*, the center's most visible publication to date. He found neither the "unity of method" nor the "general framework" one would expect from the synthetic work that Berr had promised. After quoting from a review congratulating the center for "a superb and splendidly presented synthesis that embraced all past civilizations," he wrote: "If we put aside the exaggerated compliments, what remains? What is left is just that 'historifying history,' the struggle against which, according to Mr. Berr, was the chief aim of the Center."[122] Tumenev's essay gives a sense of what Bukharin may have had in mind for Warsaw.

Bukharin's participation in the London congress turned out to be his first and last foray into academic debates on scientific history. Bukharin did, however, achieve his goal of showcasing Marxist synthesis brilliantly. The delegation he led made a lasting impression on the circle of scientists in the United Kingdom who already had an interest in Marxist approaches. The circle's biographer, Gary Werskey, aptly called it *the visible college*. It included the biologists John Desmond Bernal, J. B. S. Haldane, Joseph Needham, and Lancelot Hogben, and the mathematician Hyman Levy. On Hogben's suggestion, the Soviet delegation's papers were printed and distributed to the delegates who attended the Soviet session. Soon thereafter, they were published as *Science at the Cross Roads*.[123] The appearance of this volume helped spark interest in Marxist methodology among the intellectuals in the West.[124]

In the Anglophone world, Bukharin's paper and especially that of Boris Hessen, titled "The Social and Economic Roots of Newton's Principia," made a strong impact. Hessen's analysis of Newton's groundbreaking work as a product of its time, catering to the needs that emerged in seventeenth-century England with the rise of industry and merchant capitalism, became a classic example of Marxist historiography of science.[125] Hessen's paper directly influenced Bernal's *The Social Function of Science* (1939), *Marx and Science* (1952), *Science and Industry in the Nineteenth Century* (1953), and the three-volume *Science in History* (1954), all of which were widely read. In his monumental multivolume *Science and Civilisation in China* (published between 1954 and 2004), Needham expressed his own debt to Bukharin and Hessen. Beyond Marxists, Hessen's methodological influence extended to the classic *Science, Technology, and Society in Seventeenth-Century England*, in which Robert K. Merton reproduced Hessen's model at the same time as he distanced himself from a rigid "technological determinist" interpretation by giving centrality to the "superstructural elements" (religion and protestant ethics). By subtly substituting Hessen's "Marxist thesis" with his own "Merton thesis," he played a central role in the dissemination of Hessen's work, which became a classic point of reference in science and technology studies.[126]

Although to a lesser extent than Hessen, Bukharin too found an eager audience among left-wing scientists. In 1939, writing to Beatrice Webb, an ardent champion of the Soviet Union, Bernal reflected on his first encounter with Soviet Marxists in London: "I can say that the inspiration for my work and that of many others in science, notably Haldane, and Hogben, can be traced definitively to the visit of the Marxist scientists." He added: "We did not understand all that they said, in fact I now suspect that they did not understand it entirely themselves, but we did recognize that here was something new and with immense possibilities."[127] Bernal was not the only one who, while sympathetic to Bukharin's discussion, found it somewhat wobbly. Another scientist expressed the sentiment quite clearly. On his return from the congress, Bukharin gave a copy of his London paper to Vernadskii. After reading the paper, Vernadskii noted in his diary: "In general, [this is] interesting. But, in the upshot, in his paper 'science' is all but replaced with philosophy."[128] As a practicing scientist, Vernadskii did not recognize the science that Bukharin was talking about.

The next chapter turns to another participant in 1931 London congress, the biologist Nikolai Vavilov, Bukharin's protégé and a member of the Soviet delegation at London. Following the trajectory of Vavilov's life and work helps us see how the specific content of science entered the programs of the champions of scientific history.

3 Nikolai Vavilov, Genogeography, and History's Past Future

In the first decades of the twentieth century, the historians involved in the movement for scientific history, as we have seen, collaborated with scientists seeking to integrate studies of the past into a coherent interdisciplinary framework. They mined historical data for their research and proposed ways of using their scientific tools to shed light on the distant past, particularly where traditional historical sources were missing or ancient peoples had left few material remains and no written sources.

One such scientist was the geneticist and plant breeder Nikolai Vavilov. In the historical memory of twentieth-century science, his figure looms large. In biology, he is remembered for his pioneering work on the centers of origin of cultivated plants—the regions where a diverse range of the wild relatives of cultivated plants could be found. In the 1920s and early 1930s, he traveled around the world collecting data on plant genetic diversity and building a vast collection of seeds and plant materials that was without equal in the world at the time. This ambitious program was enabled by the alliances he had built with the Bolshevik leaders in the aftermath of the Revolution while energetically promoting genetics research as a crucial resource for the national economy, one with which to modernize Soviet agriculture. These very alliances, however, made him a target of political attacks. He is a central figure in the story of the rise to power of the agronomist Trofim Lysenko and the political persecution of genetics in the Soviet Union that culminated in 1948 when genetics was officially proscribed in the Soviet Union. A leading Soviet geneticist, Vavilov was arrested in 1940 and sent to a prison camp, where he died of starvation, while Lysenko, backed by Stalin himself, took over his institutions and his positions of power.

The story of Vavilov as a "martyr" of genetics is so powerful and so instructive that it became inseparable from the story of genetics and its political persecution in the Soviet Union that appalled the world at the time and is remembered today as one of the most powerful cautionary tales in the history of science. Yet this coupling has also produced some blind spots.

Here, I tell a less familiar story, one that reveals how Vavilov's research on genetic geography became entangled with the beginnings of the *Annales* school of historiography, which emerged as a vibrant home for scientific history in the 1930s. In his research on the origin of plant genetic diversity, Vavilov used historical and linguistic data extracted from the treatises of historians and philologists while exploring the prehistory of human settlements in regions that had left few written records. He presented his most elaborate statement on the historical implications of his research at the 1931 London congress of science historians. In the aftermath of the congress, the historians associated with the *Annales* school, such as Lucien Febvre, followed his work closely, publicizing it in the pages of the *Annales* and using it as a resource in their own work. They seemed to have shared the view that the methods of genetics held the keys to an unconventional archive complementing traditional historical sources.

Today, claims that genes can be mined for information on the deep past are commonplace.[1] In the recent *The Seven Daughters of Eve*, the geneticist and founder of the genealogy company Oxford Ancestors Bryan Sykes argued that the advent of the genomics era may mark the end of historical research as we know it: "Within the DNA is written not only our histories as individuals, but the whole history of the human race. With the aid of recent advances in genetic technology this history is now being revealed."[2] Private companies such as 23andMe translate DNA data analytics into commercialized visions of historic human migrations "written in our genes."[3] As historians discuss the limits and stakes of DNA analysis, debating whether we are indeed entering (or have already entered) a new stage of historical understanding transformed by genetics, the interactions between scholars associated with the *Annales* school of historiography and the geneticists in the Soviet

Union might be instructive, suggesting that history as we know it today has already been transformed by the science of genetics. These connections have been lost in the disciplinary memories of both biologists and historians. To recover them, this chapter reconstructs the many worlds Vavilov inhabited and the ways in which the epistemologies, material cultures, and political agendas of his project were closely intertwined and reinforced one another.

The Geographies of History and the Genetic Archive

As late as June 11, 1931, just two weeks before he boarded the plane to London along with Bukharin and other Soviet delegates heading to the Second International Congress of the History of Science and Technology, Nikolai Vavilov had no such plans.[4] A few days earlier, and unbeknownst to Vavilov, Bukharin put forward his name for inclusion in the Soviet delegation after Ivan Pavlov, the eighty-two-year-old patriarch of Russian science, excused himself, citing his old age and other commitments.[5] As a replacement for Pavlov, Bukharin proposed the forty-four-year-old Vavilov, who had been elected a full member of the Academy of Sciences two years earlier in recognition of his internationally renowned work in biology and his leadership in the field of Soviet agricultural science.[6] The combination of his international reputation and his contribution to the large-scale industrialization of Soviet agriculture in the context of the first Five-Year Plan made Vavilov a strong candidate for the delegation as the Politburo hoped to use the London congress to attract the help of bourgeois specialists and Russian scientists working abroad with the Five-Year Plan effort.[7] A secret vote by the members of the Politburo was unanimously in favor of appointing Vavilov to the delegation.[8]

As soon as he learned of the assignment from Bukharin, Vavilov quickly reorganized his plans. Determined to take full advantage of the opportunity, he extended his trip abroad to over two months, making arrangements to inspect the collections of England's oldest agricultural research institutions, Kew Gardens in London and the Institute of Crops Research in Rothamsted, and acquire seeds and plant samples. He also arranged to visit Germany's most significant center of agricultural research, the Kaiser Wilhelm Institute of Plant Breeding in Müncheberg.[9] The official purpose of the trip—attending the London congress—was not high on his list of priorities.[10] Yet he took the assignment seriously.

Vavilov's paper—he was the last speaker of the Soviet session organized on the last day of the congress—was his most elaborate statement on the implications of his research to date, which, as he modestly put it, "might be of interest to historians [and] archaeologists."[11] Contradicting most archaeologists of his time, Vavilov stated that his botanical research into the origins of cultivated crops indicated that there was not one cradle of agricultural civilization but many. He identified seven such centers: three in Asia, one in the mountainous area adjoining the Mediterranean, one in Africa, one in southern Mexico, and one in Peru. The domestication of plants and animals, he stressed, seemed to occur independently in these different locales. That is to say, the agricultures associated with these centers "have come into existence autonomously, whether simultaneously, or at different times, and one . . . must speak at least of *seven* principal cultures or, more exactly, groups of cultures." The "knowledge of the initial centers of agriculture," he suggested, can "throw light on the whole history of mankind, and the history of general culture."[12]

The view of world history that Vavilov presented in London was quite different from the way in which historians of the time thought about the geography of human history. The language he used suggests that he was well aware of the relatively new concept of the "Fertile Crescent."[13] The originator and popularizer of the idea was an American Egyptologist, James Henry Breasted, who introduced it in the two-volume *Outlines of European History* published on the eve of World War I.[14] The term *Fertile Crescent* referred to the crescent-shaped area of fertile and water-rich land that stretched from the northern littoral of the Persian Gulf up through Mesopotamia and what is now western Iran to southern Anatolia and, from there, southward along the eastern shore of the Mediterranean to the Sinai Desert. This strip of land was, the argument went, the birthplace of agriculture, villages and cities, and, hence, civilization. "Civilization," Breasted wrote, "arose in the Orient [i.e., the Fertile Crescent], and Europe obtained it there."[15]

Breasted was a well-known figure in the history of science community. He began corresponding with George Sarton in 1916; they had remained in close contact ever since.[16] In 1926, he became the second president of the History of Science Society, which Sarton had helped establish. In 1930, he published an extensive analysis of the so-called Edwin Smith Surgical Papyrus, one of the oldest survived papyri—Breasted dated it to the seventeenth century BCE—which included a medical treatise that offered researchers remarkable insights into the state of ancient Egyptian knowledge of human anatomy, diagnoses, and treatments, including surgical procedures.[17] Sarton wrote a lengthy review of Breasted's work, underscoring that it sheds new light on the perennial question, When and where did science begin? He then referred to it in his essay "East and West" (written in the same year) to argue that the ancient Middle East origins of science show the continuous link between East and West, and between the different cultures of science—those of "the old humanist and the scientist."[18]

Sarton's review came out in *Isis* in April 1931. In June, the Oxford historian of medicine Charles Singer opened the London congress with a speech that centered on Sarton's question, "When did science begin?" Singer's address singled out recent publications by Breasted and Vasilii Struve, a historian from Leningrad who planned to attend the congress but withdrew at the last moment.[19] The Edwin Smith Surgical Papyrus and the Moscow Mathematical Papyrus—the oldest "surviving papyri of scientific content"—pointed to Egypt as the birthplace of the medical, mathematical, and agricultural knowledge that was later assimilated by the Greeks and developed into the Hellenistic scientific tradition.[20]

Vavilov's data image of the centers of origin of cultivated plants was clearly at odds with Singer's imagery of the geography of history. Contradicting Singer, whose opening address he must have heard a few days earlier, Vavilov stressed: "In spite of the great historical and cultural importance of the Mediterranean centre, which has given rise to the greatest civilizations of antiquity . . . this centre, according to the investigation of its varietal diversity, includes but few autochthonically important crops."[21] Cultivated crops arrived in Egypt from elsewhere, he argued. He turned his gaze to Abyssinia, the home of the largest selection of wheat varieties in the world, a place mentioned only once and in passing in *Outlines of European History*.[22] Vavilov stated: "Though no archaeological memorials testifying to the ancient character of the Abyssinian centre have been found . . . it may be affirmed, on the basis of the diversity . . . of the plants, as well as the technique of agriculture . . . [that] still exists to some extent in Abyssinia, that this centre is indubitably independent and very ancient. It is our conviction that Egypt borrowed its crop plants from Abyssinia. All comparative data . . . point to the autonomous character of the Abyssinian centre."[23]

The Fertile Crescent concept also implied that agriculture originated in areas associated with major rivers and water-rich valleys, such as the vast basins of the Nile or in the Tigris and Euphrates river system. Vavilov argued that this was a misconception, stressing instead the overlooked importance of the mountainous regions. Taming the Nile, the Tigris, the Euphrates, and other major waterways required the building of dams and other great technological efforts to regulate seasonal flooding that "could be accomplished only by a population united into large groups," something that "could have taken place only in the later stages of the development of human society." Hence, he contended: "In opposition to the common views of the archaeologists, our investigations of the ancient agricultures have led us to the conclusion that primitive agriculture was not irrigated."[24] Instead, he argued that the first settlers developed the first small-scale agricultural technologies in the mountains, where the conditions for irrigation were simple, the climate and environment variable, and the plant life diverse. The histories of these societies, however, had been lost or distorted in world-historical narratives that focused on large settler societies and relied on their records.

In his earlier publications as well, Vavilov commented on the historical implications of his work. In *Studies on the Origin of Cultivated Plants*, published in both Russian and English in 1926, he noted that "a thorough knowledge of the cultivated plants with their multitude of . . . geographical groups" suggested that "the history of the origin of human culture, and of agriculture, is evidently more ancient than [one] narrated by such records as pyramids, inscriptions, bas-reliefs, and tombs."[25] Elsewhere, he noted that, by studying the geography of genetic variation, scholars could extend the time frame of history to encompass "distant epochs where periods of five to ten thousand years mean only a short time."[26] Plant genetics also seemed to be returning to some of history's unanswered questions about the deep history of the peoples in regions that lacked written records. Answers to these questions, Vavilov suggested, could be recovered from a different kind of record, one imprinted in the natural history of crops and cultivated plants.

Vavilov's approach represented a specific branch of genetics that developed in the Soviet Union in the 1920s under the name *genogeography*.[27] The genogeographic approach centered on organizing vast sets of data on genetic variation by placing them on geographic grids. This visual and spatial approach to the study of genetic variation developed first in agricultural genetics and then extended to animal and human genetics. After Lysenko rose to power, genogeographic studies were suppressed, as was all Soviet genetics research. In the 1970s, when genetics was officially rehabilitated, genogeography reemerged as a prominent area of research.[28] Vavilov played a central part in establishing it as a distinct approach to genetic studies.

The Mendeleev of Biology

A son of a wealthy textile manufacturer, Vavilov studied agriculture at the Moscow Agricultural Institute ("Petrovka"), from which he graduated in 1911.[29] He was one of the select few graduates recommended for a "preparation for a professorship."[30] As part of the process, he interned at the Bureau of Applied Botany in Saint Petersburg. The bureau, founded under Tsar Alexander III in the aftermath of the Great Famine of 1891–92, was part of the effort of the tsarist government to modernize Russian agriculture and prevent another food crisis. Since 1900, it had been headed by Robert Regel, the son of a German horticulturalist and botanist who moved to Russia in 1855 and assumed the directorship of the Imperial Botanical Garden in Saint Petersburg.[31] Tightly integrated within the transnational scholarly networks of his time, Regel took as the model for the bureau the US Department of Agriculture (USDA), a towering temple of American agricultural improvement and modernization in the nineteenth century.[32] Under Regel, the bureau established itself as the leading institution in plant breeding and selection in pre-Revolution Russia. Following the practices of the USDA, it assembled a large collection of the seeds of the major cultivated crops.[33] Vavilov spent several months at the bureau learning its techniques and practices and retained a close association with it thereafter.

In 1913, Vavilov departed for what he planned as a two-and-a-half-year period of study abroad, a typical path toward a professorship in Russia. He spent most of his time abroad under the tutelage of William Bateson, the chief promoter of Mendelian genetics in the years following the rediscovery of Mendel's work in 1900. He stayed for over a year at Bateson's John Innes Horticultural Institute in Merton near London, working on his dissertation on plant immunity to infectious diseases. At the outbreak of World War I, however, he had to return to Russia. He was found unfit for military service because of an old eye injury, but he was drafted anyway after Russia suffered catastrophic losses in its campaigns against German armies in February 1916. Disqualified for military service, he was discharged to the Ministry of Agriculture, which sent him to examine the causes of sickness among Russian troops stationed in northern Persia (Iran). He quickly found out that the root of the problem was bread made from diseased wheat, and, after completing the assignment, he resolved to use the opportunity to explore the region and collect disease-resistant and other valuable plant varieties. This first expedition became a turning point in his career.

The route of the expedition was dictated by wartime circumstances: to avoid a direct encounter with military action in the region, Vavilov traveled to mountainous areas in the center of the Pamir Mountains, along the border between Russia and Afghanistan. He returned with an unusually large collection of wheats, barleys, ryes, peas, and other crops. In part, this was the result of the remarkable diversity of wild crops he found there. In part, it was his unusual approach to collecting: instead of collecting the most valuable varieties—a practice adopted by plant hunters everywhere from the USDA to the Russian Bureau of Applied Botany—he collected everything. In doing so, he drew inspiration from his former mentor, Bateson.[34]

At the beginning of his career, Bateson traveled to Central Asia hoping to find invertebrate marine "relics" in the Aral Sea basin, an area regarded at the time as the surviving trace of a hypothetical prehistoric Indian Ocean.[35] Finding none, and frustrated by the apparent failure of the expedition, he abandoned his hunt for specific marine relics and set out to explore the geographic variation of invertebrates generally in the Central Asian desert. The change of fauna led to a change of methodology. Instead of meticulously dissecting individual specimens and examining them under the microscope, he began studying large sets of specimens quickly. Rather than zooming in on the features of individual organizers, he began looking for patterns in large masses of data. The new, surveying mode of research meant traveling light and getting rid of most of his fragile instruments as they were unsuited to moving around in the harsh environment of the Central Asian desert. The expedition yielded massive amounts of data on variation that took Bateson years to systematize, which he did, in particular, by arranging them into "regular series." These data formed the basis of his monumental *Materials for the Study of Variation with Especial Regard for Discontinuity in the Origin of Species* (1894), the magnum opus of his early career.[36]

Vavilov, too, spent years systematizing, cataloging, and processing the material from his first expedition. After he returned from the Pamirs expedition in spring 1917, he was offered a professorship in agriculture and plant breeding at the Agricultural Institute in the Volga city of Saratov, which he gladly accepted. He moved to Saratov in autumn 1917, just as the Bolsheviks seized power in Petrograd. Amid the calamities of the Civil War, he managed to launch and run a program of breeding experiments that analyzed the genetics of the plants he brought from Central Asia and those he found in the Volga River valley. To avoid being drowned in "the multitudinous chaos of innumerable forms" typical of traditional plant collections, he arranged variations of plant forms into "rows" and "series," just as Bateson did.[37] The result was the "law of homologous series of variation," a generalization and a numerical visualization that organized a great diversity of forms and variations within species of wild and cultivated plants in regular series. The law—the observed regularities in the series—stated that a particular variation observed in one crop would also occur in related species.[38]

From its first presentation at the opening of the All-Russian Congress of Plant Breeders in Saratov in the summer of 1920, Vavilov's law caused a stir in the Russian biological community. The president of the congress, the plant physiologist V. P. Zalenskii, reportedly exclaimed: "Biology has found its own Mendeleev!"[39] The congress adopted a resolution stating that Vavilov had "opened a new epoch" in the study of inheritance. The resolution compared the significance of his proposal for biology to the effect that the discovery of the periodic table had on chemistry: "The remarkable repeatability and periodicity [of plant forms] are opening the possibility to predict the existence of not-yet-known forms, just as the periodic table of Mendeleev allowed [chemists] to predict the existence of unknown elements." It did not stop there. The organizers telegraphed members of the Bolshevik government in Petrograd, the head of Narkompros, Anatolii Lunacharskii, and the head of the People's Commissariat of Agriculture (Narkomzem), Semyon Sereda, urging them to fund Vavilov's work "at the widest scale possible," noting that it was "a major event in world biological science, equivalent to the discoveries of Mendeleev in chemistry, one that opens broad perspectives for practitioners of the biological sciences."[40]

Analogies between the relatively new science of genetics and chemistry were common at the time. Bateson, in particular, favored analogies involving chemistry throughout his career, both before and after the rediscovery of Mendel. Thus, in 1892, discussing the work of Francis Galton (the father of eugenics), he mused that Galton's study of variation provided "an analogy . . . with the phenomena of chemical action, which is known to us as a discontinuous process, leading to the formation of a discontinuous series of bodies, and depending essentially on the discontinuity of the properties of the elementary bodies themselves."[41] Similarly, he saw the implications of Mendelian genetics for biology through a comparison with chemistry. He stressed that such traits as greenness and yellowness, hairiness and smoothness, etc. were "unit characters," susceptible, like chemical elements, to endless recombination. In his often-quoted observation: "The organism is a collection of traits. We can pull out yellowness and plug in greenness, pull out tallness and plug in dwarfness."[42]

In the first decade of the twentieth century, Bateson was an outspoken advocate for genetics (he coined the word *genetics* in 1906). Genetics, he argued, made the study of heredity look as predictable and "hard" as chemistry, epitomizing the engineering and constructivist promise of Mendelism. The notion of genes as atomistic units quickly became ubiquitous. Vavilov, who regarded Bateson as his teacher, quite likely picked up the reasoning behind the analogy between genetics and chemistry from him. Like Bateson, he envisioned the future of plant breeding as a sort of chemical engineering: "Our task is the creation of new forms. . . . In the near future we will be able to synthesize, through hybridization, the forms not known in nature. Biological synthesis is becoming a reality just like chemical synthesis."[43] For Vavilov, Mendeleev's periodic table quickly became a key reference point. Like the periodic table, Vavilov's series pointed to gaps to be filled in—plant forms that featured some combination of traits that must exist but had yet to be discovered in nature or created artificially. In an extended English-language presentation of the law of homologous series published in the *Journal of Genetics* (a journal coedited by Bateson), Vavilov stated: "It becomes possible to forecast, on this basis, the existence of forms and of varieties not yet discovered."[44]

The timing is important for placing Vavilov's law and its reception by his peers in Russia in context. In the summer of 1920, when the plant breeders assembled in Saratov, the country was verging on famine. This did not stop the Bolsheviks from regarding the countryside as a vast grain reserve they could exploit to feed the cities. One of the first decrees they adopted in 1918 was the forced requisitioning of grain, which only aggravated the famine. By early 1920s, rural areas and the cities alike were ravaged, especially in the Volga region. As members of the former elite, scientists had experienced famine firsthand. Of the forty-one full members of the Academy of Sciences in 1917, more than one-third died from disease and malnutrition during the Civil War that broke out in 1918 and lasted for almost four years. Amid death, tragedy, and deprivation, scientists struggled to continue their research.[45] By the time of the Saratov congress in the summer of 1920, the tide of the Civil War had turned in favor of the Red Army, and scientists were increasingly courting the Bolshevik leaders as their new patrons. The plant breeders thought they had something to offer. The praise Vavilov received from his peers testified not only to their trust in his ability to become a leading spokesman for agricultural science in Russia (a role compatible to the one that Mendeleev played for chemistry) but also to his ability to become a successful mediator with the Bolsheviks.

By 1920, Vavilov was widely regarded as a rising new star of agricultural science in Russia. He was the first to be offered the directorship of the Bureau of Applied Botany when Regel died from typhus in Petrograd in early 1920. Simultaneously, he was entrusted with organizing a meeting of the prestigious congress of plant breeders, which had not met since 1913. The meeting in his own city of Saratov became the rousing finale of his work there. As the congress's organizer and host, Vavilov was to speak at its opening on June 4, 1920. In what was de facto a keynote address, he sought to systematize the results he had produced since returning from Persia and the Pamirs.[46] The original title of the resulting paper was "On the Genetics of Cereals (Thoughts on the Question of the Factors of Speciation)," but sometime between May 15, when he submitted his paper proposal, and June 4, when he delivered his talk, he renamed it "The Law of Homologous Series in Hereditary Variation."[47] Fuzzy and unpresumptuous *thoughts* thus became an unambiguous and fundamental *law*. That Vavilov presented his paper at the opening of the congress may explain his choice of more ambitious terms. The stakes for agriculture and biology at this transitional moment were indeed very high.

Vavilov himself was ambivalent about what he meant by *law*. In contrast to many of his followers, he preferred to use the less rigid term *pattern* (*zakonomernost'*).[48] In 1934, when André-Georges Haudricourt, a doctoral student working with Marcel Mauss (a sociologist close to Lucien Febvre and the *Annales*), stayed at Vavilov's institute in Leningrad as an intern, he found Vavilov's fiddling with these terms somewhat annoying: "Vavilov had a favorite word, and that was the Russian word *zakonomernost'*. This is a word-for-word translation of German *Gesetzmäßigkeit* ['regularity']. In French it is translated as *une régularité tendancielle*. In fact, [he] seemed to be uncomfortable talking [about his homologous series] in terms of a 'scientific law.' And so [he] chose approximate metaphors. For me, it was almost offensive, especially this favorite term of his, *pattern*; I found it absurd as such an imprecise expression."[49]

Vavilov emphasized that the law of homologous series was a useful heuristic rather than a causal law. Arranging data on genetic variation in rows and series was a data-handling technique, a means to order vast diversity. The series became, at the same time, a blueprint for future expeditions premised on the hope of finding theoretically possible and still unknown wild relatives of cultivated crops. Vavilov's next major expedition—to Afghanistan in 1924—further confirmed his intuition about the role of the mountainous regions of Southwest Asia as hot spots of plant genetic diversity.[50] "We have found," he wrote after returning from Afghanistan in 1925, "that the mountainous regions of Asia and the mountain parts of the Mediterranean are the depositories of the varietal wealth, from which one can scoop material for [introduction] in different regions. There, we have a kind of living museum representing all the wealth of plant varieties."[51]

The future research program was clear. Parting ways with the usual itineraries of plant breeders exploring fertile river valleys, Vavilov set out to explore the world's mountain ranges, especially in Asia, in search of hidden repositories of plant genetic resources. This approach put him squarely at the center of the Bolsheviks' geostrategic interests.

Vavilov's Genogeography and the Bolsheviks' Geopolitics

Naturalists have long regarded mountains as their natural laboratories. Before the ascendancy of the Alps as a prototype for the organization of knowledge about mountains, the Central Asian mountain range was a model for naturalists seeking to validate various theories.[52] The first scholarly study of the Central Asian highlands was published by the Prussian zoologist Peter Simon Pallas, who came to Russia on the invitation of the Saint Petersburg Academy of Sciences. In 1768–74, at the request of Catherine II, Pallas traveled along the Pamir mountain range in Central Asia and elaborated one of the first theories of Asian mountain chain formation.[53] It was, however, Alexander von Humboldt who put Central Asia on the intellectual and geographic maps of European naturalists. In 1829, Tsar Nicholas I invited Humboldt to explore the Central Asia highlands. Humboldt, who dismissed Pallas's theory as a piece of "dogmatic and careless geology," set out to explore "this massif [that Pallas] vaguely called *le plateau de Tartarie*" to set an example of the proper analysis of a mountain range. To this end, he methodically measured average elevation, mapped the Central Asian mountain ranges, and provided detailed descriptions of the rocks he found there.[54] After his study, the Asian highlands quickly became one of the privileged objects for naturalists who wished to explore big questions, such as the causal relationships that connect all phenomena of the natural world.[55]

It was hardly a coincidence that the Central Asian highlands, which became so important for natural history, were also a geopolitically important region. More than many of the other mountain regions studied by naturalists, these highlands acquired a status as both a scientific and a political object.[56] For this reason alone, they attracted visitors of all sorts. Tellingly, at the same time that Bateson trekked all alone across the Central Asian desert, his British compatriots were scaling the high peaks of the Tian Shan mountain range.[57] When the tsar invited Humboldt to explore Central Asia, he did so because he understood the mountains to be both an important repository of valuable metal ores and other materials needed for industrialization and an important natural border of his expanding empire. And Humboldt understood the tsar very well. Not only did he provide a relief map of the Central Asian highlands that outlined the contours of the southern part of the Russian Empire; he also suggested a geopolitical conception of the entire region.[58] It was in this context that he coined the term *Eurasia*, which implied that the division between Asia and Europe had no basis in nature. The highlands of Central Asia constituted, in this view, a distinct region of central importance for Eurasia as a whole.[59]

In the early decades of the twentieth century, this region took on an outsized political importance as the British and Russian Empires indulged in the "Great Game," jostling for influence in Afghanistan and neighboring territories.[60] In 1904, at the height of the distrust that verged on open war between the two empires, Halford John Mackinder, an Oxford geography professor and the director of the London School of Economics, argued that the Eurasian continent had become the strategic center of what he called the "World Island" (comprising Asia, Africa, and Europe) as a result of the economic and industrial development of southern Siberia and the decline of the European overseas empires. He considered that "crescent of marginal land"—the Central Asian highlands bordering the central part of the Eurasian continent—the "pivot of history" because, to conquer the vast expanse of Eurasia, one would have to gain control of it.[61]

Twenty years later, in 1924, mindful of the lessons of World War I and the collapse of the Russian Empire, Mackinder revised his thesis and prophesied that the power of the Eurasian heartland could be offset by Western Europe and North America.[62] The Bolsheviks, who had entered the Great Game as its newest player, thought otherwise. In the aftermath of the October Revolution, the Bolshevik strategists—from Lenin to the head of the People's Commissariat of Foreign Affairs (Narkominotdel), Georgii Chicherin—believed that the revolution in Russia had triggered a revolutionary movement everywhere in the world. They considered Persia, for instance, key to closing off access to India, "the citadel of the revolutionary East."[63] Neighboring Afghanistan was to their minds even more important since it provided the closest overland routes to India.[64]

Vavilov's expedition to Afghanistan in 1924 coincided with the Bolsheviks' efforts to establish relations with Tehran using diplomatic, commercial, and scientific channels all at once. The Bolsheviks' geopolitical goals in the region not only made Vavilov's expedition possible but also shaped the expedition's agenda and Vavilov's own work. The expedition, which Chicherin personally oversaw, was funded by the Narkomzem and the Narkominotdel.[65] Vavilov was issued a diplomatic passport that presented him as a "plenipotentiary diplomatic representative of the Soviet Union." The two researchers who accompanied him carried official papers presenting them as diplomatic couriers of the Narkominotdel.[66] Vavilov not only explored all the territory of Afghanistan but was also eager to collect any information he could that would be valuable to the government. For instance, as he recalled, he secretly photographed the British installations in the restricted zone between Afghanistan and India, disguising the intelligence gathering as plant collecting.[67] After returning from this expedition, he was awarded one of the first Lenin Prizes, the highest Soviet award honoring exceptional service to the country.

From his first expedition to Persia and the Pamirs in 1916, Vavilov proved himself a skilled diplomat. And, as a skilled diplomat, he left few traces of the intense networking that enabled his collecting missions. In his publications, he always downplayed politics. In his travel journals, which he distilled into the unfinished *Five Continents*, he would occasionally remark on "getting mixed up in diplomacy," but always ironically and in passing.[68] The extensive study of agrarian Afghanistan that he coauthored with his research associate and published as *Zemledel'cheskii Afganistan* included 318 photographs and maps of the region and was praised as "a serious contribution not only to agricultural knowledge but also to Eastern studies."[69] Yet, as one reviewer noted, "the great shortcoming of the book" was that it omitted "any information or discussion of the political situation in the region, which, as the authors themselves underscored, functions as a strategical buffer between India and Russia."[70]

The traces of Vavilov's intense and mostly secretive networking at the highest political levels surface in unexpected places. In 1926, for example, Vavilov was involved in the negotiations around the mission to Tibet pursued by the painter and occultist leader Nikolai Roerich, who spent much of the 1920s and 1930s engaged in political schemes mixed with espionage and self-delusion. After emigrating from the Soviet Union in 1917, Roerich and his wife, Elena, ventured on a covert, occult-inspired mission that they referred to as the "Great Plan": an aspiration to create a Pan-Buddhist state, Shambhala, that would unify Asia and include most of Central Asia, parts of Siberia with largely Muslim populations (Altai, Tuva, and Buryatia), Mongolia, western China, and Tibet. As their plan progressed, the Roerichs started to seek alliances with the Bolsheviks who aspired to unify Muslims across Asia under the banner of communism.[71]

Vavilov knew Roerich quite well. In the 1920s, they collaborated on various projects. In 1921, they participated in designing the Russian pavilion for the exhibition "America's Making" (held at the Seventy-First Regiment Armory in New York), which showcased the contribution of immigrants to American life.[72] The same year, Vavilov established the New York office of the Bureau of Applied Botany, hiring Roerich's follower and confidant, the entomologist Dmitrii Nikolaevich Borodin, as a managing director (Borodin acted as a messenger between Vavilov and Roerich).[73] When Roerich came to Moscow in 1926 to negotiate a deal with the Bolsheviks, Vavilov was involved in the negotiations. He tried to use Roerich's contacts and the Bolsheviks' interest in Tibet to realize his long-held desire to explore the region where he believed was to be found the oldest surviving form of ancient high-altitude agriculture. As part of the negotiations, he wrote a letter to the Dalai Lama in Lhasa requesting that the "Most-Learned and Glorious Ruler of Tibet" assist him in obtaining samples of seeds of cultivated plants from high-altitude mountainous regions of his country. In return, he offered to supply the seeds of plants cultivated in the Soviet Union or any other country of the world from the rich collections at his institute.[74]

During the 1920s, Vavilov aligned his program with the goals of the Bolshevik elite and became a part of the ruling elite himself. Although not a Communist Party member, as a member of the Central Executive Committee of the Soviet Union (TsIK, the highest Soviet legislative body), he participated in Party congresses and conferences.[75] He had the ear of the highest-ranking Bolshevik officials, such as Bukharin and Aleksei Rykov (the chairman of the Sovnarkom from 1924 to 1930).

Vavilov succeeded in building alliances with the Bolsheviks to support his expeditions partly because he did not need to convince them of their importance. There was in Russia an established tradition of exploratory expeditions carrying out extensive (and expensive) studies of the country's territory, its flora and fauna, its peoples, and its boundaries. In the aftermath of the Revolution, the number of expeditions skyrocketed. The Soviet government not only supported scientists like Vavilov but also organized scientific expeditions on its own. Nikolai Gorbunov, a personal secretary of Lenin's who became the secretary of the Sovnarkom after Lenin's death, organized a number of geographic, botanical, anthropological, and other sorts of expeditions to the Pamirs, Mongolia, Africa, and other parts of the world.[76] During the 1920s, Gorbunov, educated as a chemist-technologist, was the chief patron of leading Soviet scientists and one of Vavilov's main patrons among the Bolsheviks. He not only supported Vavilov but also served as a research associate, chairing the scientific council of Vavilov's Institute of Applied Botany and New Crops (later renamed the Institute of Plant Industry), which replaced the old Bureau of Applied Botany. An avid traveler and explorer himself, he personally participated in several expeditions, including a collaborative Soviet-German expedition to the Pamirs in 1928 during which he collected plant specimens for Vavilov.[77]

Vavilov's ambitious program would not have been possible without support from high-ranking Bolshevik officials. When the financial backing fell short (as it often did: Vavilov frequently lamented the lack of funds at his disposal in comparison to the level of funding available to plant breeders at the USDA), generous allotments of manpower, buildings, and land for experimental stations made up the difference. With these provisions, Vavilov was able to reorganize the small-scale Bureau of Applied Botany as the much larger and more formidable Institute of Plant Industry in Leningrad, which employed several hundred researchers. In addition to the institute, Vavilov established a network of some one hundred experimental testing stations in different climates and regions all over the Soviet Union.[78] After Afghanistan, he led expeditions to the Mediterranean, Abyssinia (Ethiopia), Eritrea, and Yemen (in 1926–27), Bavaria (1927), western China, Japan, and Taiwan (in 1929), and Central and South America (1932–33). When he could not travel himself, he sent his research associates to, for instance, Central Asia in 1925 and Central America in 1930. By the late 1930s, the number of staff working at the institute approached one thousand, and the collection of seeds amounted to some 200,000 specimens from all over the world. At the time, it was the largest and most systematic such collection in the world.[79] Foreign scientists who visited the Soviet Union in the 1920s invariably had Vavilov's institute on their travel agenda (see chapter 4 below).

Vavilov's institute, nicknamed "Vavilon" (meaning "Babylon" in Russian) for its grandeur, became the standard on which international agricultural research organizations in the later twentieth century would model their gene banks.[80] Biologists who visited the institute in the 1920s and 1930s were captivated by its scale. William Bateson, who visited in 1925, used superlatives in the hasty notes in his travel diary: "All going very well. Experiences of the most varied. The rush is really more than normal U.S.A. . . . Today is 'Science Holiday' throughout Russia! Vavilov's place on enormous scale: says, some 400 people, of whom about half are [scientifically] trained. Has seeds of 13,000 varieties of wheat etc in perfection."[81]

Collecting plants and seeds from everywhere, taking measurements and cataloging the data, storing the seeds, rendering them accessible for analysis, and regularly planting and then harvesting them to maintain vigor and viability is difficult enough. But collecting was only the beginning. Along with working on the development of high-yield varieties of major crops, Vavilov processed and blended different types of data into a uniform whole. If the law of homologous series provided a blueprint for future expeditions, his second generalization—the centers of origin of cultivated plants—provided a data image of the plant genetic resources of the entire planet, identifying the sites of major concentrations of plant genetic diversity. The data image, not surprisingly, bore the imprint of his travel itineraries.

The entanglements of scientific and geopolitical interests not only enabled Vavilov's collecting missions; they also shaped the resulting data image of the hot spots of biodiversity. Over time, Vavilov changed the number of centers identified: from three in 1924, to five in 1926, six in 1929, seven in 1931, and eight in 1935, each addition a result of a consideration of new data from later expeditions.[82] The mountainous region of Central Asia, which he initially identified as a "primary" center, was subsequently demoted to merely a "secondary" one, an area rich in biodiversity through human intervention and, thus, more recent than any of the primary centers. He was consistent, however, in identifying the Asian highlands, where he traveled the most extensively, as the richest repository of valuable crop varieties.[83] Notwithstanding such wavering, his mapping of plant genetic diversity provided a road map for future studies of genetic resources. In the 1930s, the United States and Nazi Germany both sent expeditions to the regions that Vavilov identified as centers of genetic diversity; and both established national collections like the one in Vavilov's institute.[84] His expeditions, his institute, his collection, and his identification of centers of plant genetic diversity made Vavilov an inspirational figure for breeders and plant collectors the world over.[85]

When Bukharin suggested that Vavilov be included in the delegation attending the London congress, he made a shrewd calculation. Not only was Vavilov recognized as one of Russia's most renowned biologists; he was also a sincere socialist and a devoted patriot, acting as an ambassador of Soviet science both publicly and privately. On June 11, 1931, just a couple of weeks before flying to London, he wrote to Theodosius Dobzhansky, a geneticist who had left Russia in 1928 on a Rockefeller fellowship and was working in Thomas Hunt Morgan's genetics laboratory at the California Institute of Technology, hoping to convince him to return to Russia: "I have no doubt that despite all the calamities it's much more interesting to live in the Soviet country. Life might be more comfortable in California for certain individuals, but, on the whole, our way is more just, and you know this."[86] Years later, Dobzhansky would describe Vavilov as a "passionate Russian patriot" who was often "considered a Communist, which he was not." "But," Dobzhansky continued, "he accepted the Revolution wholeheartedly because he thought it opened greater possibilities for Russia's people. . . . He asserted that nowhere else in the world was the work of the scientist valued as highly as in the USSR."[87]

While not a Marxist like other members of the Bukharin-led delegation to the London congress, Vavilov was genuinely sympathetic to dialectical materialism as a philosophy of science that emphasized the unity of knowledge.[88] His own program, after all, sought to unify knowledge across disciplinary divides. As he put it in *Five Continents*, the thread connecting all his work was an aspiration "to unify what was perceived as difficult to connect—geography, botany, agronomy, [and] the history of culture."[89] In his letter to Dobzhansky, he cited the law of homologous series as an example of the dialectical method. Dismissing Dobzhansky's concerns about the imposition of Marxist theory on all aspects of intellectual life in the Soviet Union, he wrote: "Dialectical methodology is but a plus [in science]."[90]

In the late 1920s, the genetic-geographic approach that Vavilov developed working with plants was extended to animals. The geneticist Aleksandr Serebrovskii, the founder of the Department of Genetics at Moscow University, christened this approach *genogeography*—the mapping of "the geographic distribution of the concentration of various genes."[91] To many Soviet geneticists, genogeography was primarily a tool with which to organize and visualize data.[92] Some, however, shared Vavilov's more expansive vision. Serebrovskii, for instance, echoed Vavilov when he stated: "The current geography of genes is the result of a lengthy historical process, and, hence, by studying the geographic distribution of the various genes, we will be able to compile a detailed history not only of domestic livestock but also of man."[93]

Traditionally, biologists doing the kind of work Vavilov was doing—charting the geographic distribution of plant diversity—relied on historical studies as repositories of useful data. One of the founders of botanical geography, the nineteenth-century Swiss botanist Alphonse de Candolle, the son of another botanical geographer, Augustin Pyramus de Candolle, mined the treatises of historians and philologists to extract information about the variability and geography of cultivated plants that he then used to complement his botanical-geographic studies.[94] And, indeed, Vavilov used de Candolle's studies as an inspiration and a source for his own research, dedicating his own 1926 *Studies on the Origin of Cultivated Plants* to "the memory of Alphonse de Candolle . . . as a tribute to his '*L'origine des plantes cultivée*,' among his other classical works." However, he pointed out that de Candolle relied too much on "archaeological, historical and linguistic . . . ancillary methods," which he "applied without [substantial] botanical basis."[95] Moreover, de Candolle was a synthesizer who relied on the works of others, rarely leaving his study.[96] "De Candolle's classical work," Vavilov wrote, "occupies still an exceptional position. [Yet] the somewhat scanty information of A. de Candolle has been considerably enlarged by new historical and archaeological data, and by series of botanical and genetical discoveries."[97] As Vavilov traveled around the world and analyzed his findings, he came to believe that, as he put it, "many historical problems can be understood only because of the interaction between man, animals, and plants."[98] At the London congress, Vavilov set out to convince the attendees of just that.

A "New Kind of History"

Arguing that "the botanic-geographical analysis of cultivated plants" was enabling historians to find answers to some of the unanswered "questions connected to the history of general human culture," Vavilov may sound prescient today.[99] Some of his contemporaries, such as the British biologist J. B. S. Haldane, saw his work as science "of the future" as well as a glimpse into the "future of history," an exciting new form of historical research, as Haldane argued in an essay provocatively titled "Is History a Fraud?" published in 1930 in *Harper's Magazine*. Haldane questioned professional historians' fundamental assumption that, without written documents, there is no history (see chapter 1). He reasoned that the reliance on written records alone left many people without history at all: "Historians have inevitably thought in terms of words. They have read many books and documents. They have often been great stylists like Gibbon and Macaulay. They have realized the power of words to move magnitudes." Yet, he continued: "[They] have seldom realized that man's hands are as important as and more specifically human than his mouth. . . . When I look at history, I see it as man's attempt to solve the practical problem of living. The men who did the most to solve it were not those who thought about it, or talked about it, or impressed their contemporaries, but those who silently and efficiently got on with their work."[100]

Haldane argued that Vavilov's work was an example of such a "history of the future" already in place, enabled by the Marxist dialectical framework and the alliance of science and socialism in the Soviet Union. In 1928, he visited the Soviet Union at the invitation of Vavilov, who had arranged for him to stay at his institute in Leningrad. Inspired by what he saw, Haldane prophesized that, in the future, a new "form of historical research" would be conducted by scholars who would make the separation of written history from prehistory, human history from geohistory, and historians from scientists obsolete. Adopting a "bird's eye view of history," and exploiting synergies between historical knowing and scientific investigation, these future scholars would give a history to the peoples who did not have it in the standard historical narratives.[101]

Haldane was not alone in seeing Vavilov's work as a new form of historical research. In 1929, when Lucien Febvre and Marc Bloch launched the *Annales d'histoire économique et sociale*, they envisioned a "new kind of history" that would reconcile human history with other studies of the human (and nonhuman) past (see chapter 1 above). Soon after the 1931 London congress, French scholars associated with the *Annales* school began to follow Vavilov's work closely. Febvre's patron, Henri Berr, had, as we have seen, attended the congress along with a few other members of the International Center of Synthesis, which he and Febvre codirected. Even though, after founding the *Annales*, his connection with the center weakened, Febvre continued to be in close contact with Berr.[102]

Another channel through which Vavilov's presentation at the London congress could have reached French historians was the Cercle de la Russie neuve, a French socialist organization with close ties to the All-Union Society for Cultural Relations with Foreign Countries (Vsesoiuznoe obshchestvo kul'turnoi sviazi s zagranitsei, better known by its acronym, VOKS), a Soviet organization established in Moscow in 1925 to coordinate contact with intellectuals and cultural figures in the West. The Cercle de la Russie neuve, which positioned itself as an intellectuals' organization, was established soon thereafter. Between 1931 and 1933, it organized several discussions of *Science at the Cross Roads* (the published proceedings of the Soviet session at the London congress), publicizing the book among French intellectuals as "one of the most important published documents on the tendencies and achievements of science in the Soviet Union."[103]

The historian of agriculture Charles Parain, one of the cofounders of the Cercle de la Russie neuve and a close associate of the *Annales* group, played a crucial role in acquainting French historians with Vavilov's work. Parain championed Marxist history and what he characterized as a "materialist conception of the history." In his own work, he pointed to indigenous knowledge of plants and nonindustrial, agrarian techniques as alternative points of departure for the historians of modern agriculture.[104] In his historical reconstructions, he built on Vavilov's work and publicized it in the *Annales* throughout the 1930s.[105] Febvre, in turn, publicized Parain's work, stressing that the question of the origin of agriculture had been "reopened by the recent work at the Institute of Applied Botany in Leningrad, particularly by Vavilov." He highlighted, in particular, Vavilov's conclusion that "irrigated valleys could not have been the original centers of agricultural techniques." In consequence: "To attribute the first great anonymous agricultural victories to the Egyptians and the Chaldeans would be mistaken."[106]

Another close associate of Febvre's, the sociologist and ethnologist Marcel Mauss (a nephew of Émile Durkheim's), was also interested in Vavilov's work. In 1929, Mauss was a participant in the first week of synthesis at the International Center of Synthesis, which featured three conjoint conferences: one of scientists discussing biological evolution, one of historians discussing "civilization, the word and the idea," and one of the historians of science holding their first international congress (see chapter 1 above).[107] In his presentation, Mauss attacked Oswald Spengler's much-discussed *Decline of the West* as an example of how *not* to do a scientific history of civilizations: "Spengler's 'morphology of civilizations' is, in our opinion, a literary [rather than a scientific] enterprise. . . . This philosophy of history . . . has no value except as a piece of [cheap] popularization." At that time, Mauss doubted that the biological sciences could inform what he called a *cultural anthropology of civilizations*, an approach that emphasized local and unique features of different cultures. He might have had the discussions of evolution at the parallel sessions in mind when he stated: "[Cultural anthropologists] oppose especially—although perhaps too easily—the simplistic ideas that present human evolution as if it was a universal process."[108]

Genetics, however, presented a different ontology. As Staffan Müller-Wille argued, the combinatory logic of genetics created room for historical contingency. The cultural anthropologist Claude Lévi-Strauss, in particular, used contemporary genetics as a resource to overcome the linear and deterministic view of history inherent in evolutionary accounts of humanity.[109] Before Lévi-Strauss's interest in genetics spiked in the 1950s in the context of his involvement in drafting the UNESCO statement on race, Mauss, whose work Lévi-Strauss considered as formative for his own, started to look in this direction himself.[110]

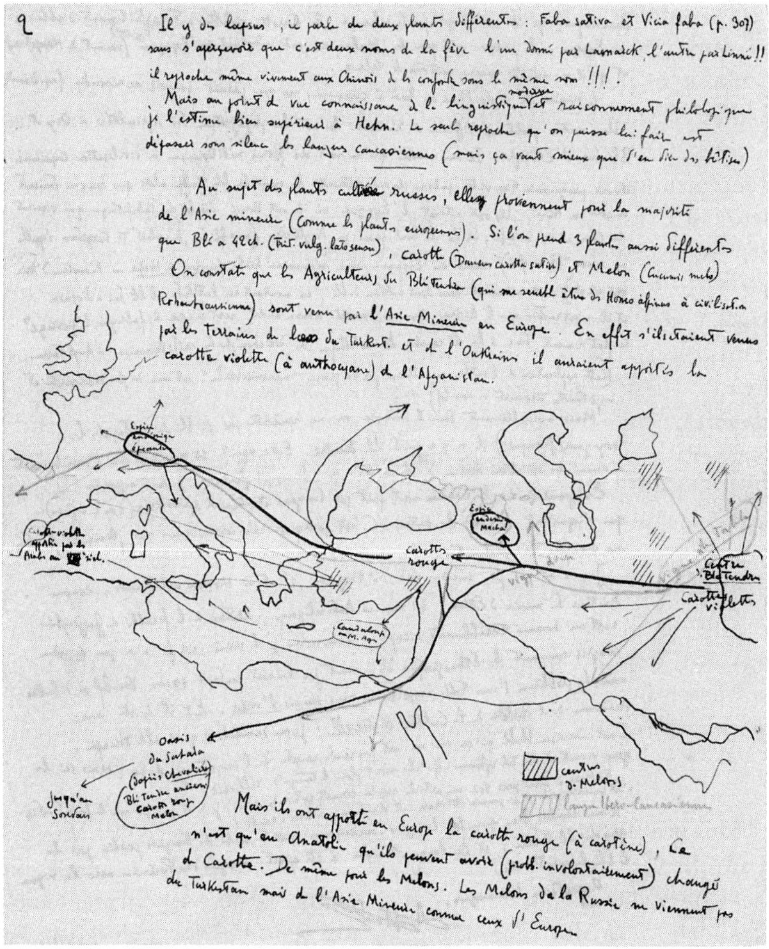

FIGURE 2 A drawing of Vavilov's centers of origin in a letter from A.-G. Haudricourt to Marcel Mauss, 1935. Image courtesy of the Collège de France.

In 1934, Mauss arranged for his student André-Georges Haudricourt to intern at Vavilov's institute in Leningrad, securing financial support for his stay through his own Ethnology Institute in Paris.[111] Throughout his year-long internship, Haudricourt, who had a background in linguistics and agronomy, kept up an extensive correspondence with Mauss, describing Vavilov's genogeographic methods at length (fig. 2).[112] After his return from Leningrad, he proceeded to develop what he referred to as "la dialectique géographique" and present it at the meetings of the Marxist study group.[113]

The interest of French scholars in Vavilov's genogeography cannot, of course, be explained simply by individual events such as the 1931 London congress that popularized Vavilov's work among historians. Rather, Vavilov's work attracted the *Annales* scholars because they had already been engaged with ongoing research in biology, an engagement stimulated by the weeks of synthesis and other initiatives of the International Center of Synthesis. Specific conclusions about the implications of his research, as Vavilov presented them at the London congress, resonated with the very credo of the *Annales* school of historiography, which placed the natural world and geographic space in the foreground of historical analysis.[114] In the paradigmatic example of the Annaliste approach, *The Mediterranean and the Mediterranean World in the Age of Philip II*, Fernand Braudel echoed, in particular, Vavilov's observation that agrarian practices in the mountainous highlands had not left the kinds of records historians normally use and could be recovered only by using unconventional archives: "The Mediterranean is, above all, a sea ringed round by mountains. This outstanding fact and its many consequences have received too little attention in the past from historians. . . . The history of the mountains is checkered and difficult to trace. Coming down from the mountain regions, where history is lost in the mist, man enters in the plains and towns the domain of classified archives."[115] None of Vavilov's publications appear in Braudel's bibliography, but the imprint of his work is easily traced through the copious citations of Parain's works in *The Mediterranean and the Mediterranean World*.

This is not to say, however, that Braudel was influenced by Vavilov in any direct or causal way. Rather, we can discern similar emphases on problems and questions that historians of the *Annales* school and biologists such as Vavilov shared. In a similar vein, Vavilov's work can be compared methodologically with Lucien Febvre's landmark *Geographical Introduction to History* (1925) without suggesting that this work could have influenced Vavilov's own thinking about geography and history, which was most certainly not the case. Back in the Soviet Union, however, his notion of history, and its alleged connection to Western scholarship, had become a matter of public discussion and put Vavilov in a hot spot.

The Politics of History

By the time that Vavilov was picked by Bukharin to represent Soviet science at the 1931 London congress, he had already begun to experience the beginning of the troubles to come. As recently as 1929, he had been at the peak of his power. In that year, he organized the First All-Union Congress of Genetics, Seed Production, and Plant and Animal Breeding in Leningrad, designed to highlight the achievement of Soviet genetics. The congress, organized with the support of the Soviet government, was the pinnacle of his personal success as a recognized spokesman of Soviet genetics and agricultural science and made the headlines of leading Soviet newspapers. Gorbunov, then a member of TsIK and of the Narkomzem's executive board, gave him the ultimate endorsement by writing a lead article for *Leningradskaia pravda* highlighting the importance of his work for the modernization of Soviet agriculture.[116] In the aftermath of the congress, Vavilov himself was appointed a member of the Narkomzem executive board, a sign that the value of his effort to modernize Soviet agriculture was recognized.

The year 1929 was, however, also a turning point. The beginning of the Great Break marked the beginning of Vavilov's downfall. Vavilov experienced the first challenge to his authority within his own Institute of Plant Industry when, in the immediate aftermath of the genetics congress, the institute's political sector (*partiinoe buro*) protested that its demand to include "Marxist forces" on the congress's organizing committee had not been met. Between 1930 and 1931, the conflict grew into an organized internal campaign against Vavilov. He was criticized for, among other things, his "unnecessary fascination with expeditions" and his "authoritarian methods" of running the institute.[117] The conflict quickly took on political overtones in the context of the show trials of the so-called Industrial Party, which came on the heels of the Academic affair (see chapter 2 above) and targeted agronomists, many of whom were close colleagues of Vavilov's, on fabricated charges.[118] Amid the gathering clouds, on March 11, 1930, the Soviet secret police, the NKVD, began conducting covert surveillance on Vavilov.[119] The same year, his main patron among the Bolsheviks, Nikolai Gorbunov, was removed from his position as a secretary of the Sovnarkom.[120]

At this point, however, the signs of Vavilov's imminent downfall were still hidden from public view. The first public attack appeared in print in 1932 when the Academy of the History of Material Culture (Gosudarstvennaia akademiia istorii material'noi kul'tury, or GAIMK) published a paper by one of its members, the archaeologist and specialist in the ancient history of Central Asia Georgii Grigor'ev, titled "On the Question of the Centers of Origin of Cultivated Plants."[121] In it, Grigor'ev discussed Vavilov's major publications, including *Studies on the Origin of Cultivated Plants* and *Agrarian Afghanistan*, and claimed that his work lent support to "migrationist theories" of cultural change—a framework widely accepted among Western archaeologists at the time that explained cultural change by positing that cultures originated in a small number of centers and then dispersed through migration and diffusion.[122] This assault on Vavilov and his theories took on heightened importance in the context of the treacherous political environment of the early 1930s as manifested within the GAIMK.

The GAIMK was a remarkable institution. A brainchild of Vladimir Lenin, Mikhail Pokrovskii, and the Marxist linguist Nikolai Marr, it was established in 1919 by decree as a flagship institution aimed at reinventing archaeology as a Marxist synthetic science of antiquity. Pokrovskii suggested calling it the "Academy of Material Culture"; Lenin added "History"; Marr headed the new institution as its president.[123] Unlike the Communist Academy, created around the same time to develop a Marxist counterpart to the bourgeois humanities, the GAIMK put less emphasis on the Marxist theoretical framework and focused instead on a unifying framework of the history of material cultures that bridged the human sciences, such as archaeology, linguistics, and ethnography, and the natural sciences, such as geology, paleontology, and botany. It emphasized collaborative research across the natural sciences and the humanities, bringing different disciplinary methods to the study of ancient material remains. It also had close connections with Vernadskii's Commission on the History of Knowledge (KIZ; see chapter 2 above). Many members of the KIZ, including the historian Vasilii Struve, a corresponding member of the International Committee of the History of Science, were affiliated with the GAIMK as well.[124]

Vavilov's work was widely known within the GAIMK. Ethnologists and archaeologists referred to his expeditions as a model to emulate. In 1931, for instance, the head of the GAIMK's Institute for the Study of the People of the Soviet Union, N. M. Matorin, cited those expeditions as a model of a "complex" and "genuinely Marxist" approach that "contributed to the prestige of the Soviet Union" in the political as well as the scientific spheres.[125] The GAIMK's president, Nikolai Marr, knew Vavilov quite well, both professionally and personally. They both lived in the House of the Academicians, an elite residential building in the center of Saint Petersburg where a score of Russia's most renowned scientists continued to live with their families after the Revolution.[126] As was common among the inhabitants of the house, Marr and Vavilov had regular conversations about matters of common interest.[127] In his work, Marr vehemently attacked Indo-European comparative linguistics, arguing that, rooted in nineteenth-century German philology, it was racist at its core. He argued that "there is no racially separate Indo-

European family" and that "originally there existed not one but a plurality of tribal languages" that developed into contemporary languages through mixing, hybridization, and the formative influence of the social environment.[128] He proposed the so-called Japhetic theory of the evolution of languages, which identified languages with classes and environments rather than nations or races, as a Marxist alternative to bourgeois comparative linguistics.[129] Privately, as Vavilov noted in his diary, Marr saw Vavilov's work as lending support to his theories, which endorsed genetic concepts of hybridity and crossbreeding in relation to languages and the evolution of world cultures.[130]

Vavilov himself, however, was agnostic with regard to any causal explanation of cultural change, the main question that migration theories sought to provide an explanation for. There is no evidence that he was even aware of Grigor'ev's publication. The assault from this unlikely corner was characteristic of the atmosphere of the Great Break, which turned the academic discussions within the GAIMK and elsewhere in the Soviet Union into acerbic political attacks. In the midst of the Academic affair, many archaeologists at the GAIMK lost their jobs. Some of them rebranded themselves as *geologists*.[131] Grigor'ev had chosen a different strategy, turning himself into a militant Marxist.[132] In his paper, he claimed that Vavilov's mapping of the centers of origin of agriculture implied an anti-Marxist understanding of the evolution of world cultures. Moreover, he claimed that the migrationist theory to which Vavilov's work has allegedly lent support was racist at its core and had originated within the tradition of comparative Indo-European linguistics, condemned by Marr. All this placed Vavilov on the wrong side: "Vavilov endorses this Indo-European 'theory' of migrations, which is thoroughly deceitful and chauvinistic." He explained further: "Vavilov's error is that he shares the perspective of Indo-European linguistics, . . . a reactionary and chauvinist West-European linguistic theory [that] derives Indo-Germans from some Indo-Germanic *Herrenvolk*."[133] By lending support to migrationist theories, Vavilov, he claimed, supported assumptions about an "Indo-Germanic 'master race'" and, by extension, the rising German fascist discourse about the supremacy of the Aryan race.[134]

Even though Grigor'ev's paper does not seem to have had a significant role in Vavilov's downfall, it is telling that the first public assault on Vavilov came from a historian, not a biologist and that it targeted his (alleged) notion of history, not his biology. By 1935, a massive Lysenko-led assault targeted Vavilov and genetics. At the meeting of the Soviet Agricultural Academy (Vsesoiuznaia akademiia sel'skokhoziaistvennykh nauk imeni Lenina, or VASKhNIL) in Odessa in June that year, Lysenko, in particular, targeted the law of homologous series, criticizing Vavilov for an overly simplistic conception of heredity, and arguing that the law became an arbitrary standard for the selection and breeding work at Vavilov's institute. Lysenko's supporters also attacked Vavilov's vast seed collection, arguing that it had made no contribution to Soviet collective farms.[135] Then, in 1938, Lysenko took over the VASKhNIL—an institution originally created and presided over by Vavilov. Two years later, Vavilov was arrested and subjected to brutal interrogation.[136] He died of starvation in 1943 in a prison in Saratov, the town where he began his meteoric career.

Vavilov's tragic fate, the details of which would not become known in the West until the end of World War II, became a symbol of the calamities of Soviet genetics. In the meanwhile, the links between Vavilov's genogeography and the developments in historiography around the *Annales* school in the 1930s had been obliterated in the disciplinary histories of both biology and historiography. Paradoxically, the memory of the *Annales* project as a revolution in the writing of history and that of Nikolai Vavilov as a martyr of genetics have reproduced the very disciplinary divisions that the *Annales* historians and biologists such as Vavilov called into question.[137] As the next chapter shows, however, Vavilov's tragic fate and the crisis in Soviet biology became the backdrop against which another participant in the London congress, the biologist Julian Huxley, launched his own campaign for a scientific history.

4 Julian Huxley's Cold Wars

In January 1949, soon after stepping down as the director-general of UNESCO, the biologist Julian Huxley wrote a memo to his successor on what he regarded as one of the organization's most important projects, an ambitious plan to write a new "Scientific and Cultural History" of humankind. "I regard this as a key project for Unesco," Huxley stressed. "A scientifically based history of mankind as a whole, conceived from the cultural and scientific angle, is necessary to provide a common ground [fulfilling] Unesco['s] central task of promoting the advance of a world civilisation. It could become in some sort the Bible of the nascent world civilisation to which we look forward."[1]

Under UNESCO's auspices, scientific history would be propelled to an unprecedented level of political support, resources, and visibility. None of this would have happened without the administrative support that Julian Huxley was able to marshal as UNESCO's first director. To understand how this somewhat unlikely figure came to champion scientific history as a path to world peace, this chapter traces Huxley's trajectory between 1931, when he helped organize the Second International Congress of the History of Science and Technology in London, and the late 1940s, when, at the onset of the Cold War, he launched the UNESCO history project, positioning it as an answer to the unfolding political crisis.

Huxley's advocacy of scientific history was, this chapter argues, critically shaped by his experiences, first, as a political traveler in the Soviet Union in the aftermath of the 1931 London congress to observe the Soviet experiment in action and, then, as a biologist actively following the Lysenko-led campaign against genetics in the Soviet Union. In August 1948, Huxley participated in the Soviet-sponsored World Peace Congress in Wrocław, Poland, which took place soon after the fateful events in the Soviet Union that effectively criminalized genetics in that country. When he first traveled to the Soviet Union in 1931, he thought that Soviet "scientific socialism" could provide useful lessons for Western societies. Now, he was convinced that the Soviet antiscience position with regard to genetics was a dangerous ideology that threatened the intellectual culture of the West. Concurrently engaged in the development of the evolutionary synthesis and in political battles with Soviet ideologues, Huxley turned evolution into a general worldview that would, he hoped, provide the West with an all-encompassing framework to counter Soviet universalist ideology.

At the time of 1931 London congress, Huxley identified himself as a "scientific humanist," someone who refused to split existence into two realms, that of science and that of human affairs. His second career as an author by trade provided him with a platform from which to promote a scientific or evolutionary history as an intrinsic part of scientific or evolutionary humanism, which he positioned as the West's counterpart to Marxism.

Julian Huxley's Two Careers

When Charles Singer took charge of organizing the Second International Congress of the History of Science and Technology, he invited Huxley to join the congress's advisory council, which featured British scientific celebrities such as the physicists Ernst Rutherford and Arthur Eddington and included the heads of Britain's most important scientific institutions, who lent their names to support the congress as a vehicle to promote science and scientific education through the history of science. Huxley did not represent any scientific institution, appearing simply as "Prof. Julian Huxley of London" in the congress program.[2] Unlike other spokesmen for science in interwar Britain, who communicated their work to the general public while maintaining active research careers, he eventually abandoned his first career as a professional scientist to devote himself full-time to writing about biology and human affairs.[3]

A grandson of Thomas Henry Huxley's, Darwin's famous "bulldog," and on his mother's side the great-nephew of the Victorian cultural critic and poet Matthew Arnold, Julian Sorell Huxley was born into the British intellectual elite. (As he put it in his memoirs: "I was born with great advantages, genetic and cultural.")[4] With an excellent education, received, according to the standards of his social status, at Eton and Oxford, he began what promised to be an illustrious academic career. At the age of twenty-two, he was appointed a lecturer and demonstrator in the Department of Zoology and Comparative Anatomy at Oxford. Shortly thereafter, he published his first scientific paper—on regeneration in sponges—in the prestigious *Philosophical Transactions of the Royal Society*, "a real honor for a biologist only twenty-three-years old," as he noted in his memoirs.[5] That same year, he published his first book in the genre he dubbed *philosophical biology*. Titled *The Individual in the Animal Kingdom*, it was a reflection on the notion of biological individuality inspired by the French philosopher Henri Bergson.[6] Soon thereafter, the president of the newly founded Rice Institute in Houston, Texas (today's Rice University), offered him a professorship. In 1912, at the age of twenty-five, he moved to the United States to build Rice's new Biology Department.

The outbreak of World War I interrupted Huxley's meteoric rise, but not for too long. After enlisting into the army and spending the last two years of the war in Italy as a British army intelligence officer, Huxley took a lectureship position at Oxford, which had lost some of its leading faculty and was actively recruiting its former students as replacements.[7] He published several important papers and played an important part in the establishment of experimental zoology in Britain as a cofounder of the Society for Experimental Biology (SEB) and the *British Journal of Experimental Biology* (*BJEB*) in 1923.[8] In 1925, King's College London offered him a chair in biology with a salary of £1,000 a year, the highest academic salary in the country at the time.[9] Soon after assuming the chair, however, he resigned, accepting an offer from the popular author of "scientific romances" H. G. Wells to collaborate on writing a book on biology for a broad audience. He never again held an academic position, turning his writing on science and human affairs into a full-time job.

From the very beginning of his academic career, Huxley was interested in and published prolifically on broad questions relating to the role of science in the modern age, the relationship between science and religion, the meaning science could give to individual lives, and the social roles of scientists. In 1923, the same year that he cofounded the SEB and the *BJEB*, he published his first book for general audience, *Essays of a Biologist*, a collection of articles he had written for various journals. Another collection, *Essays in Popular Science*, followed in 1926. The next year, in *Religion without Revelation*, Huxley presented his philosophical credo, arguing that traditional religion no longer suited the modern world and that a new, universal religion was needed, one that reconciled scientific rationality with secular spirituality.[10] He would articulate this perspective again and again in numerous books, pamphlets, and papers, advocating a view that science—and biology in particular—provided not just specific knowledge but a general framework explaining human existence and fulfilling deep spiritual, psychic, and social needs. In his view, this science-based religion would become a secular counterpart to traditional religion.[11]

Huxley's writings exemplified a specific discourse on science as a secular substitute for religion that was common in Britain in the interwar years. A year before Huxley published *Religion without Revelation*, a younger contemporary, the budding embryologist Joseph Needham, edited a collection of essays, *Science, Religion, and Reality*, that explored similar themes. Besides Needham, several leading British scientists and scholars representing different fields contributed essays on the subject.[12] In his own, mildly quizzical *The Great Amphibium* (1931), Needham sought to convince skeptics "to admit that there may be a religious aspect in science."[13] Like Huxley and many other contemporary scientists, he wove together biology, politics, religion, and philosophy in pursuit of a unified and comprehensive account of the world.[14]

The notion of science as a modern substitute for religion was popular among the nascent community of historians of science as well. George Sarton (see chapter 1 above) gave this discourse on science a name—the *new* or *scientific humanism*, where *new* stood for secular and scientific. As a committed Comtean positivist, Sarton saw the decline of religion as a mark of scientific progress. He opened *The History of Science and the New Humanism* by outlining his "faith of a humanist" and arguing that, as "the consciousness of mankind," science was to replace religion and superstition as the ontological expression of human existence.[15]

Singer was one of the leading advocates of scientific humanism in Britain. He and Huxley knew each other from their days at Oxford, where both had studied zoology and later been recruited as lecturers in zoology and the history of medicine, respectively.[16] In 1921, he moved to London to accept a better position in the new Department of History and Method of Science at the University College, London, but he and Huxley continued to stay in close contact. In 1922, when he coorganized an Oxford summer school on the history of the interconnections between science and civilization, he invited Huxley to contribute.[17]

While Sarton and Singer aimed their gospel of scientific humanism at academics of all stripes, Huxley sought to reach a wider audience, one beyond the academy. In 1929, he launched a magazine, *The Realist: A Journal of Scientific Humanism*, whose intended readership was "the bank clerk, . . . the trade union official, the small professional man." He invited Singer to contribute an agenda-setting essay for the opening issue. Singer made the case for the history of science as the embodiment of scientific humanism. "A survey of science conveyed through history should aid [a student] in understanding why his life is worth the living," he argued, insisting that the history of science should be taught as a central part of history.[18]

In the early 1930s, Huxley and other contributors to *The Realist* started to use the new medium of radio to reach out more directly to their intended audience, those "small professional men" who were, as it turned out, reluctant readers of intellectual journals. As the historian Robert Bud has argued, it was radio that made scientific humanism a part of the national conversation in Britain. This medium implied a more conversational tone and accessible rhetoric, and Huxley, for one, not only toned down his discussion of scientific humanism but also made it entertaining. As he explained the gospel of scientific humanism to his listeners in one of his talks: "My point is that [in scientific humanism] you get a general outlook . . . which would be scientific all around, just as you have a nationalistic outlook, or a religious outlook, or a socialist outlook."[19] He quickly became a regular speaker on the BBC, his voice eventually reaching hundreds of thousands of listeners curious about the relationship between science, religion, and society.

The way for the success of Huxley's radio talks had been prepared by his involvement with the publishing industry in the nascent field of science journalism.[20] With a reputation for being a scientist who could write for the general public, Huxley was regularly commissioned by various popular magazines and daily newspapers. When he took up his professorship at King's College in 1925, the move to London opened up new possibilities for his literary activities, beginning with his collaboration with H. G. Wells.[21]

Best known as a novelist and science fiction writer, Wells also tried his hand at history. His best-selling *The Outline of History* (1920) integrated human history within the larger story of the origins of the earth and the universe, seeking to reconcile the developments in the physical and biological sciences with those in archaeology, linguistics, and historiography. Written in collaboration with large number of scholars and scientists, *The Outline of History* was intended as a "plain history" for a general audience rather than a scholarly work.[22] After the book sold in the hundreds of thousands and earned him the handsome sum of £60,000, Wells began looking for someone to collaborate on a biological sequel.[23] He invited Huxley to coauthor a similarly wide survey of biology, *The Science of Life*. Given the success of *The Outline of History*, the publisher George Newnes offered Huxley a large advance and estimated a remarkable profit of at least £10,000 (a decade's worth of Huxley's salary at King's College). Huxley accepted Wells's proposal without much hesitation, asking only that the project be put off long enough for him to prepare his exit from the academy.[24]

Huxley resigned from King's College in 1927. *The Science of Life*, cowritten with Wells and his son Gip (G. P. Wells), who had trained as a biologist at Cambridge, was produced with great speed and published in thirty-one fortnightly parts between March 1929 and May 1930.[25] Even though the sales of *The Science of Life* were lower than those of *The Outline of History* had been, they exceeded the expectations of the publisher, and the textbook version continued to sell well into the 1950s and 1960s. The collaboration with Wells became a turning point in Huxley's life. Abandoning his academic carrier, Huxley turned his literary and public activities into a full-time job.

By the time of the 1931 London congress of science historians, Huxley was a household name and one of the best-known scientists in Britain.[26] His daily calendar for the week of the congress, from June 30 to July 4, 1931, is packed each day with appointments that must have kept him away from the sessions.[27] Apparently, he did not even make it to the symposium on the relationship between physics and biology that Joseph Needham had organized or to the Soviet session on the last day.[28] But he was active behind the scenes. In particular, he met the members of the Soviet delegation whose surprise arrival caused such a stir in the London press (see chapter 2 above).

Huxley might have been more aware than most of his British colleagues as to the composition of Soviet delegation and its high-profile status (which came as a surprise to many of the congress participants), thanks to his contacts with the science journalist James G. Crowther. A dropout from Cambridge, where he read mathematics, Crowther was the in-house science correspondent for the left-wing *Manchester Guardian*, the first British newspaper to create a staff position for a correspondent covering science news.[29] A devout Marxist, he was keenly interested in the implications of socialism for science. In 1929, he traveled to the

Soviet Union for the first time; he visited again the following year and produced a book-length report, *Science in Soviet Russia*.[30] In early 1931, working with his connections in the Soviet Union, he began organizing a group tour to the Soviet Union for British scientists. The idea was to establish formal contacts between British scientists and their Soviet colleagues.[31] In early spring, Huxley had signed up for one of Crowther's tours that was to depart right after the congress, and he had been busy preparing since May.[32] During the congress (at which Crowther acted as a guide for Bukharin), he met Nikolai Vavilov and Boris Hessen. They would meet again later in the summer in Leningrad.[33]

Huxley had both a political and a literary interest in joining Crowther's tour of the Soviet Union. On the political side, he was involved in debates about planning that had unfolded in Britain during the great economic depression in Europe and the United States after the stock market crash of 1929. Several years of unprecedented social and political instability had fueled a sense of insecurity and fears that another economic meltdown would jeopardize the entire social and political order. Amid what was widely perceived as a crisis of old-style market capitalism, an interest in the Soviet experience with planning spiked. In February 1931, Huxley joined the group Political and Economic Planning (PEP), which had been organized by Max Nicholson, a zoologist and regular contributor to the Tory magazine the *Week-End Review*. PEP sought to develop and promote a version of "planned capitalism" that would reconcile the techniques of large-scale socialist planning with a private capitalist economy.[34]

A trip to the Soviet Union also presented an excellent publishing opportunity for a writer with Huxley's reputation. His brother, the writer Aldous Huxley, also planned to join the trip, for much the same reason.[35] During the 1920s and the early 1930s, a trip to the Soviet Union became a rite of passage for Western intellectuals of all political stripes.[36] The pilgrims to "new Russia" included noted literati—the Briton H. G. Wells, the Frenchman André Gide, and the German Lion Feuchtwanger, among many others—all of whom wrote up dispatches from their political travels.[37] Wells made more money from *Russia in the Shadows*, in which he described his personal encounter with "the Kremlin dreamer" Vladimir Lenin, than he did from *The Outline of History*.[38] Among numerous travelogues in the genre of "back from the Soviet Union," however, Crowther's *Science in Soviet Russia* (1930) was the only English-language book on Soviet science. There was a clear demand for more.

Huxley clinched a lucrative contract with Harper and Brothers before he set sail on a Soviet ship. Besides Crowther and Huxley, the group included Huxley's wife Juliette (née Baillot), the writer and translator Ivy Low (who was married to the Soviet diplomat Maxim Litvinov), and several British medical doctors.[39] Huxley spent three weeks in the Soviet Union, mostly in Leningrad and Moscow. The day after his arrival in Leningrad, he went to see Vavilov and the Institute of Plant Industry (see chapter 3 above).[40]

A Journey to a Utopian Future

In his book about the journey, *A Scientist among the Soviets*, Huxley mused about being dazzled both by Vavilov's vast collection of seeds and by his vision of the future. His descriptions of the visit are full of didactic dialogues. For example, he described how he invited Vavilov's comments on the Five-Year Plan by musing quizzically: "While the Five Year Plan is without doubt of the greatest importance, it is in a sense only an incident. . . . For when, after two more years, it comes to an end, it is to be succeeded by a Ten Year Plan. . . . And undoubtedly this will be succeeded by other plans in due course." Vavilov took the bait: "You will see . . . a Hundred Year Plan for Russian agriculture, and if it would be extended to world agriculture, perhaps a Thousand Year Plan." Walking around the institute's field station, housed in a Victorian-style palace, Huxley expressed his surprise: "How Victorian!" Vavilov readily explained: "You are perfectly correct. The house was given, complete, by Queen Victoria as a present to one of the Russian Grand Dukes in 1887." What emerged was a vivid image of the palaces once occupied by Russian aristocrats turned into the palaces of science, the lords and vassals of the tsarist regime having moved out, and the scientific labs having moved in. Huxley did not comment on the scene, but his point was crystal clear: scientists enjoyed support and recognition in the socialist utopia.[41]

Some version of these conversations did indeed take place. Thanking Vavilov in a letter after the visit, Huxley expressed his concern that his host might have "overstrained [his] voice telling us so many interesting things."[42] The tone of Huxley's published account, however, departed dramatically from that of his hastily written notes taken during the journey (fig. 3). What he saw of Russia and the "achievements" of socialism sincerely shocked him. He registered the atrocious living conditions—"the many empty stores, queues, . . . the absence of cafes and restaurants"—and mocked "propaganda, slogans" everywhere. On the whole, he found Russia "terribly depressing."[43]

His travel notes suggest that at first Huxley struggled to find a theme

FIGURE 3 Julian Huxley on his tour of the Soviet Union, July 1931. Image courtesy of the Woodson Research Center, Rice University.

that would bind his observations into a didactic story. After a few days, however, he scribbled in his travel diary: "[I began to see] what I saw [as] . . . a utopian future state of humanity."[44] A "futuristic" perception of new Russia became the guiding theme of *A Scientist among the Soviets*. "Russia today," Huxley mused, "is a transition between a medieval past and a communist future, a compromise between chaos and a plan, a mixture of expedience and principle. Over and over again you see evidences of this hybridity." Using a biological metaphor, he compared "Soviet Russia in the third year of its Five Year Plan" to "an embryo halfway between conception and birth."[45]

Considering the timing of Huxley's trip helps contextualize *A Scientist among the Soviets*. By the summer of 1931, the economic crisis in Europe and North America had reached its high point. In August, the financial crisis led to the collapse of the Labour administration in Britain and the formation of the so-called National Government of August–October 1931, an emergency coalition designed to rescue the country from complete collapse. Huxley embarked on his journey right in the middle of this turmoil. As he described it in *A Scientist among the Soviets*: "We left England for Russia just when the German financial crisis of July 1931 was engaging the world's attention. . . . Then, after three weeks, we returned—to find not only Germany and England, not only western Europe, not only the industrial nations, but the whole capitalist world, in the throes of a crisis, a crisis which since then until this time of writing has grown steadily more acute."[46]

On the Soviet side, in contrast, Huxley's travel coincided with the retreat from the Great Break policy of terror directed at the old professional intelligentsia (see chapter 2 above). In July, the official rhetoric swung from an emphasis on the "last battle" between old and new Russia to an emphasis on positive incentives to attract technical specialists, both old and new, in Russia and from abroad. The Soviet tourist agency Intourist, created in 1929 to replace VOKS (the All-Union Society for Cultural Relations with Foreign Countries), actively began to promote a particular vision of the Soviet Union to Western visitors. In 1931, Intourist launched an advertising campaign under the slogan "a journey to the new world."[47] A tour to the Soviet Union was advertised as a trip through time, a trip to the future. Consciously or not, Huxley reproduced precisely the imagery that Intourist used to sell the less than perfect conditions of the country to foreign visitors.

It has often been noted that travel narratives blend the seen and the imagined.[48] Huxley took the imagined to the extreme. *A Scientist among the Soviets* blended observation and imagination, exploiting explicit comparisons with fiction. Describing an encounter with a "German town-planning expert" who traveled all over Siberia looking for sites "where cities are to arise" and imagining "the broad outlines of the future city," he mused: "A modern Arabian Nights, with a modern genie on a modern version of a magic carpet." This account, full of similes and metaphors, made the project of engineering utopias as undertaken by omnipotent states and omniscient intellectuals such as Vavilov seem plausible. More than anything else, Huxley projected an image of the new Russia as a place where the old dream of a knowledge-based society was being revitalized. As he put it in *A Scientist among the Soviets*: "The elevation of science and scientific method to its proper place in affairs [puts] new Russia, even in its present embryo stage of development, . . . in advance of other countries."[49]

Despite the fact that it was unambiguously pro-Soviet, *A Scientist among the Soviets* was never translated into Russian. This was somewhat unusual. Accounts by foreign political travelers visiting the Soviet

Union, including Wells's *Russia in the Shadows*, were tracked down and translated into Russian.[50] Western authors were often invited to visit the Soviet Union, the expectation being that a travel narrative would result from the trip.[51] In Britain and the United States, on the other hand, *A Scientist among the Soviets* was a moderate success at best. A few reviews noted Huxley's admiration for the Soviet commitment to state-funded scientific research, but overall the response was lukewarm.[52] As a narrative that blended fiction and reality, the book might have appeared of limited use to those who sought information on actual Soviet experiences with planning, large-scale farming, and the organization of science.

Huxley's trip to the Soviet Union did, however, established his reputation as a commentator on Soviet affairs in general and Soviet science in particular, and his voice became an authoritative one within PEP and other important political groups. At the end of 1931, soon after his return from the Soviet Union, Nicholson formed a study group within PEP called "Tec Plan"—shorthand for "Technique of Planning"—that set out to elaborate a theory of planning to provide PEP with a general framework for developing a national plan for Britain and, possibly, a Planning Party. Huxley was intimately involved with the development of the Tec Plan.[53] He was also actively involved in another, more right-wing group, The Next Five Years, which included some Tories, some progressives, and a few socialists. Both Tec Plan and The Next Five Years aimed at developing a new interpretation of planning that would transcend political divisions through planned capitalism.

As Huxley's involvement in the debates about planned capitalism might suggest, his interest in large-scale planning had more to do with scale than it did with socialism. In 1934, while on a lecture tour in the United States, Huxley went to see the dams along the Tennessee River and studied the general organization of the Tennessee Valley Authority (TVA). For Huxley, the TVA, set up during the New Deal era as a federally owned corporation designed to bring electric power to the rural American South, was an exemplar of large-scale planning implemented in a region nearly the size of England. His enthusiastic account of the project, *TVA: Adventure in Planning* (1943), was the first book about the TVA written by a foreign observer.[54] It took Huxley nearly a decade to write the account, and in the end it omitted the central theme entertained in earlier publications on the subject: the comparison of the TVA with large-scale planning in the Soviet Union.[55]

In the early 1930s, Huxley's comparisons between Soviet and American large-scale planning projects fit into an established narrative. During this time, many Western European intellectuals brought together the Soviet Union and the United States under the common denominator of "the great dream of modernity" driven by faith in the world-transforming power of science and technology. The differences between these two systems notwithstanding, many of Huxley's contemporaries regarded American and Soviet projects of industrial modernization as different but not necessarily contrary realizations of the dream.[56]

By the end of decade, however, a new political imaginary of the Soviet Union—one emphasizing the differences, not the similarities, between American capitalism and Soviet socialism—began to be used for polemical effect. Between 1936 and 1938, the leading Bolsheviks who had played significant roles in the October Revolution stood show trials and were executed. The fate of one of them, Nikolai Bukharin (see chapter 2 above), inspired the journalist and political traveler to the Soviet Union Arthur Koestler's powerful political thriller *Darkness at Noon*,[57] "the most devastating exposure of Stalinist methods ever written."[58] Later historians would call this period the Great Purges or the Great Terror.[59] During the worst period of the Great Terror, many people, scientists including, were arrested, tortured, and killed. As part of the unraveling political purges, the unfolding crisis in Soviet genetics became a key element in the new political imagery of the Soviet Union in the West. It was Huxley, among the others, who wrote the new script for how the West would perceive science in the Soviet Union in the years leading to World War II and the onset of the Cold War.

The Crisis in Soviet Genetics and Julian Huxley's Cold Wars

When Huxley visited Vavilov and the Institute of Plant Industry in 1931, he followed in the footsteps of a number of Western geneticists who had traveled to the Soviet Union not necessarily because it was a destination for political travelers of all stripes but because it was an important world center of genetics research. Leading foreign geneticists such as William Bateson, Erwin Baur, and Richard Goldschmidt visited the Soviet Union in the 1920s and toured the genetics laboratories and institutes in Moscow and Leningrad. J. B. S. Haldane visited Vavilov in 1928, as we have seen, and wrote enthusiastically about the state of biology in the Soviet Union (see chapter 3 above). After returning from the Soviet Union, Haldane wrote in an essay published in *The Realist*: "There is an enormous interest in Russia in biology, because evolution is part of the Communist faith. To take one example, in the whole world the only quantitative work on natural selection . . . is being done in Russia, mostly on crop plants. . . . In the scientific study of animal heredity Russia is ahead of the rest of the world except the United States, and I think that in another ten years it will probably be ahead of the United States."[60]

Haldane's comment was perhaps tongue-in-cheek given that Soviet geneticists had been given a boost by their contacts with the American geneticists T. H. Morgan and his students (nicknamed the "fly boys"), who turned a fruit fly into a model organism of genetics. In 1922, one of the "boys," Hermann Muller, Huxley's former protégé at Rice Institute, went to the Soviet Union, bringing with him a collection of standardized mutant strains of *Drosophila*, thus introducing genetic methods developed by Morgan's group.[61] The connections had remained tight since that time. Soviet geneticists worked at Morgan's lab, and Morgan's student Calvin Bridges spent six months working at Vavilov's institute in the winter of 1931–32. In 1932, Muller accepted Vavilov's invitation to head a genetics lab at the Institute of Plant Industry. He moved to the Soviet Union with his family in September 1933.[62]

As did many geneticists at the time, Huxley included, Muller linked the insights of genetics to eugenic visions of perfecting the human race, which resonated with the Soviet vision of creating a "new society" and a "new man."[63] When Muller moved to the Soviet Union, he was convinced that only a socialist society could seriously pursue the most ambitious programs in genetics and its applied branch, eugenics. This view was shared by many geneticists at that time regardless of their political views or their position vis-à-vis socialism. In 1936, a leading American eugenicist, Charles B. Davenport, the director of the Eugenics Record Office at Cold Spring Harbor, reached out to the US secretary of state with a warning: "I have told many students of human genetics of the United States that Russia is taking the lead away from the United States in this subject, which it formerly held."[64] Muller, too, thought that the best way to proceed was to win the heads of the government to the cause, so he reached out to Stalin. On May 5, 1936, he sent him his *Out of the Night*, a collection of essays in which he advanced his colorful proposals about "positive eugenics," which included such measures as the artificial insemination of female volunteers with the sperm of "gifted individuals" (his examples of such individuals included Lenin, Newton, Pasteur, Omar Hayyam, Pushkin, and Marx).[65] In the letter accompanying the book, he urged Stalin to implement such measures in the Soviet Union as an opportunity to improve the genetic makeup and, consequently, the well-being of his nation.[66] Stalin never responded, but, as Muller put it in a letter to Huxley, he learned: "[Stalin] has been reading the book, is displeased with it, and has ordered [an] attack against it."[67] The "attack," as he made it clear to Huxley, was part of the larger campaign against genetics that erupted in earnest around that time.

The campaign against genetics played out against the backdrop of the deterioration of relations between the Soviet Union and Nazi Germany, which in the fall of 1936 became openly hostile after the Soviet Union supported the Republicans in the Spanish Civil War while Germany backed Franco. An intense antifascist campaign erupted in the Soviet press in the following months, lumping genetics together with fascism, Nazi racial cleansing, and racism in general. Leading Soviet geneticists were attacked and condemned as supporters of eugenics.[68] During the Moscow trials, which began in January 1937, some geneticists were arrested on charges of spying for Germany. The Seventh International Congress of Genetics, scheduled to be held in Moscow in 1937, was canceled by Politburo, which rescinded its earlier approval.[69]

By the time the congress had been canceled, Muller had witnessed Lysenko's increasing attacks on Vavilov firsthand. Soon, as the tide of accusations rose, Vavilov advised Muller to leave the Soviet Union, warning him that it was no longer safe for him to stay there. In February 1937, Muller fled to Spain to join the fight against fascism in the Spanish Civil War. On his way there, he sent a long letter to Huxley, detailing the unfolding campaign against genetics in the Soviet Union. According to Muller, Vavilov's connections to Bukharin and other politicians condemned as "enemies of the people" during the Moscow trials "helped to get [the congress] in disrepute."[70]

The furious reaction of scientists and journalists in the West apparently made the Politburo reverse its decision and tentatively agree to host the genetics congress in 1938. By then, however, the organizing committee had already settled on Edinburgh as a backup venue, nominating Vavilov as the congress's president. Just before the congress finally opened on August 23, 1939, the organizers received a telegram from Vavilov informing them that he and the roughly fifty Soviet geneticists who had planned to attend were withdrawing.[71]

The telegram was the last communication Vavilov's colleagues in the West received from him. On the same day the congress convened its first plenary panel, the Soviet Union signed a nonaggression treaty with Germany; a few days after the congress ended, Germany invaded Poland, and World War II began. The channels of communication between Soviet geneticists and their colleagues in Britain and the United States were severed during the war. With no information available, many suspected that Vavilov was no longer alive.

In 1942, the Royal Society made a calculated move, electing Vavilov as its foreign member. Besides paying tribute to Vavilov and sending a signal to the Soviet authorities, the election provided a rationale for an official inquiry into Vavilov's fate. After all, the society did not grant membership to deceased individuals.[72] In an open letter published in 1948, the society's president, Henry Dale, stated that "repeated inquiries" had been sent to the Soviet Academy of Sciences "through all available channels, asking only the date and the place of [Vavilov's] death," but that "no reply of any kind" had been received.[73] By that point, however, the fact of Vavilov's death had been confirmed by Huxley.

Huxley remained as active as ever throughout the war. Amid the tumult of the air raids in wartime London, he wrote *Democracy Marches* (1941), in which he proposed the reform of the British social service system. He also finished his scientific magnum opus, *Evolution: The Modern Synthesis*.[74] Published in 1942, *The Modern Synthesis* was almost immediately recognized as a landmark. In the following years, it was internationally hailed as one of the founding texts of what became known as *the evolutionary synthesis*. In 1943, when the Council of Allied Ministers of Education began considering plans for a new postwar international organization, the United Nations, the British Ministry of Education recruited Huxley as a spokesman for science within the organization. In this role, Huxley participated in 1945 and 1946 in a series of intergovernment conferences that led to the establishment of the United Nations and its scientific branch, UNESCO. In June 1945, when delegates from fifty nations assembled in San Francisco to draft the UN charter, he was among several scientists gathered in Moscow to discuss and lobby support for UNESCO.

Taking a plane to Leningrad and, from there, a train to Moscow, Huxley traveled as an "honorary guest" of the Soviet Union who had been invited to participate in the celebration of the 220th anniversary of the Russian Academy of Sciences. This was a high-profile event, with a number of foreign scientists attending on Stalin's suggestion.[75] Huxley had another urgent item on his agenda besides networking on behalf of the planned international agency: he wanted to find out what was happening with Soviet genetics. Raissa Berg, a geneticist who accompanied him in Leningrad during his visit, recalled him asking her for contact information for the geneticists who had survived the purges and the war.[76] During his stay, Huxley met with some of them in Moscow and Leningrad. From these contacts, and from Eric Ashby, a British botanist who stayed in the Soviet Union for a year in 1944–45 as a scientific attaché, he confirmed what many in Britain and the United States had already long suspected. Vavilov had been arrested sometime in 1940 and died in prison sometime between then and 1943.[77]

Huxley also wanted to use the occasion to meet Lysenko in person. He tried to arrange for an interview through his hosts at the Academy of Sciences but was informed that Lysenko was not available.[78] He soon learned, however, that Lysenko would be giving a lecture as part of the official program of the academy's celebrations. Language would not be a barrier: Lysenko had arranged for a simultaneous translation for "foreign guests," that is, Huxley and Ashby.[79] The lecture was to be held at the Institute of Genetics in Moscow, which Vavilov had directed from 1934 and until his arrest in 1940. Lysenko took over the institute in 1941.

Huxley's archive keeps an English-language summary of the main points of Lysenko's lecture and a transcription of the question and answer session that followed. It was likely provided at the request of Huxley and Ashby, who were clearly awed by what they heard and wanted to have it on record. They received a remarkable document. Lysenko opened the lecture with the following statement: "There are two interpretations of heredity: the formal Mendelian-Morgan interpretation and the modern Lysenko-Michurin interpretation. These interpretations are totally opposed to each other and cannot be conciliated." He then criticized Mendelian genetics for its "mechanistic" and "formal" approach to heredity, which reduced organisms to combinations of genes that no one could see: "Some say the heredity is in the chromosomes. Some say it is in the plasma. It does not matter to us where it is. These are all metaphysical postulates and they have nothing to do with life." Explaining his own interpretation of heredity, Lysenko stated: "Our theory is diametrically opposite. Heredity is only from the body we can feel. There is no organ of heredity. There is no hereditary matter separate from the soma. There are organs of reproduction but no organs of heredity. Heredity is the property of living matter to demand its own conditions and to demand its own food." He accused Mendelian geneticists of introducing a "tendency in science [that] makes things more complicated than they are" and stated that he and his followers "can with greater ease [than the geneticists] solve the laws of living organisms."[80]

During the question and answer session, Huxley asked: "How does Lysenko explain recombination and segregation if chromosomes play no part in heredity?" Lysenko answered: "This phenomenon cannot be explained satisfactorily on Mendelian explanation. But we have an explanation for it. . . . In fertilization one cell assimilates the other cell, or the cells mutually assimilate each other. In the conflict to decide which cell will assimilate the other, which will win? The variety which assimilates more effectively will be the variety which appears in the next generation. The segregation which sometimes appears in the F2 is simply the belching out again of the material assimilated by the stronger variety."[81]

The experience left Huxley stunned. He commented on his encounter with Lysenko in his memoirs: "It was interesting, though maddening, to see a real fanatic, a Savonarola of science, in operation."[82] He was also shocked by the scale of the devastation that the war had wrought and the "great deal of hardship" he observed over his two-week stay in the Soviet Union.[83] Intourist menus featuring various delicacies and the lavish feasts arranged for the foreign guests during the celebrations of the academy's jubilee only accentuated the poverty and devastation visible on the streets and related by fellow geneticists.[84]

Huxley did not mention any of this in the published account of the visit, however. Instead, he produced an upbeat portrayal of the flourishing state of Soviet science in the fields of evolutionary biology and genetics.[85] As Nikolai Krementsov explained, the strategy adopted by Western geneticists at this time was to use all available sources to support Soviet geneticists while refraining from any direct critique of Lysenko or any political comments that might provoke the Soviet authorities to retaliate.[86] Moreover, Huxley was also involved in the negotiations leading to the organization of UNESCO, in which the Soviet Union, one of the founding members of the United Nations, was expected to take part as well. UNESCO was inaugurated in November 1946 in London, with Huxley appointed as its first director-general.[87] The mission of UNESCO was to champion and protect international peace, which called for compromise, not confrontation.

National and international events would soon force geneticists to reevaluate this strategy. The relationship between the wartime allies deteriorated rapidly in 1946 and 1947. Churchill's speech of 1946 about the "iron curtain" falling down and dividing Europe marked the symbolic beginning of the Cold War. In January 1947, George Kennan articulated the doctrine of containment as a road map for American foreign policy.[88] In September, Andrey Zhdanov, appointed by Stalin to direct Soviet cultural policy and work with foreign Communists, delivered a speech that depicted the postwar world as "irrevocably divided into two hostile camps" representing two opposed world systems.[89] The two-camps mind-set underpinned the establishment of the Communist Information Bureau (Cominform) soon thereafter. In April 1948, following the Communist coup in Czechoslovakia, the United States launched Marshall

Plan. In June, the Soviet army blocked access to Allied-controlled sectors of Berlin, starting the first major crisis of the Cold War and prompting an unprecedented humanitarian campaign: a massive American and British airlift of food and supplies to the besieged city.

On August 19, 1948, as the Berlin crisis was making headlines around the world, the *New York Times* published a lead article reporting that Lysenko's doctrine had been declared the official stance of Soviet biology and that the teaching of genetics had been effectively banned in the Soviet Union. In the following days, the *Wall Street Journal*, the *Washington Post*, and the *Los Angeles Times* featured stories on "the crisis in Soviet biology."[90] For scientists in North America, Britain, and Western Europe, that crisis was as troubling as the Berlin crisis.

At the time the news broke, Huxley was in the midst of a different crisis. He had been scheduled to attend the World Congress of Intellectuals in Defense of Peace in Wrocław in the last week of August 1948. The congress had been organized by the Polish publisher and Francophile Jerzy Borejsza, who intended to "turn the Congress into a *bombe atomique de la paix*."[91] Knowing that the congress was sponsored by the Soviet Union and the Soviet-backed Cominform, Huxley agreed to participate with some hesitation and as a private citizen rather than as the director of UNESCO.

The Wrocław congress attracted almost five hundred leading artists, writers, scientists, and academics from all over the world. Huxley was a star guest, elected one of the congress's five presidents. Instead of becoming, as some attendees had hoped, the seed for a Soviet-led alternative to UNESCO (which the Soviet Union did not join), the parade of the world's intellectual celebrities in Soviet-controlled Poland became one of the defining moments of the Cold War.[92] The Soviet delegates soon began attacking Western literary and scientific culture, denouncing Western intellectuals as "war mongers" (Alexander Fadeev compared T. S. Eliot and Jean-Paul Sartre to "hyenas and jackals").[93] Huxley protested and left before the congress ended, refusing to sign an anti-West resolution, and denouncing the congress as an incitement to hatred.[94]

Huxley's performance at the Wrocław congress made headlines in the Soviet press. The leading newspapers, including *Pravda*, *Izvestia*, and the *Literaturnaia gazeta*, mocked him and denounced him as a *pacifist* (the word was used pejoratively in the Soviet Union) whose talk of scientific humanism distracted from the real business of the "struggle for peace in all the world." "Apparently," wrote a critic in the *Literaturnaia gazeta*, "Huxley thought that he was participating in a literary *salon* with the literati speculating about 'cultural' and 'intellectual' topics, or playing the 'games of minds,' only to realize that he was among people really willing to fight for peace, progress, and democracy."[95] The journalist David Zaslavskii, infamous for his denunciations of the poets Osip Mandelstam and Boris Pasternak and the composer Dmitrii Shostakovich in the 1930s, wrote in *Pravda* that "Huxley and his enthusiasts" instilled the congress with "a language of common phrases and 'lofty' pointless chatter." He added: "Under the banner of the defense of 'Western culture' [Huxley] was preaching . . . the cosmopolitan depersonalization of national cultures."[96]

Apparently, Huxley had finally had enough. After his premature departure from Wrocław, he toured Eastern Europe while following reports on the unfolding crisis in Soviet biology. As soon as he returned from the trip, he wrote an article on Lysenko and the situation in Soviet genetics for *Nature*.[97] While he was working on the article, he received an offer of a book contract on the topic from Henry Schuman, a New York–based publisher. He accepted, initially planning simply to expand his article.[98] In the resulting book, however, he shifted his focus from explaining why Lysenko's claims were inconsistent with science—the focus of the *Nature* article—to the more general implications of the Lysenko controversy for the Western world. "The scientific aspects of the controversy," Huxley explained, "are subsidiary to the major issue of the freedom and unity of science."[99]

Huxley came to believe that the rise of Lysenko and the tragic fate of Soviet genetics were symptoms of a larger phenomenon: the folly of a technologically rooted party-state in which an antiscience position became a political platform. When he first traveled to the Soviet Union in 1931, he thought that the Soviet model of a strong state guided by scientifically minded intellectuals like himself could provide a useful model for Western societies, regardless of the ideological differences. Now, he was convinced that the Soviet Union had become an ideological zealot threatening the intellectual culture of the West. In *Heredity, East and West* (1949), he argued that, instead of debating Lysenko, "what men of science [should] do" was develop "an equally powerful" ideology of "equally general appeal" as that of communism and Marxism-Leninism, whose all-encompassing, totalitarian nature made the phenomenon of Lysenko possible.[100]

In the late 1940s, many Western intellectuals who earlier shared a fascination with the Soviet revolutionary project used their literary prowess to advance an intellectual counterattack against communism. In 1947–48, a group of ex-Communists who had traveled to the Soviet Union in the early 1930s assembled a volume titled *The God That Failed* that powerfully expressed their disillusionment with Communist ideology.[101] Around the same time, intellectuals in the United States (and, to a lesser degree, Britain and France) began to link together fascism and Soviet communism under the common denominator *totalitarianism*. Hannah Arendt's *The Origins of Totalitarianism* (1951) gave the term its political currency.[102] Theorists of totalitarianism such as Arendt lumped Nazism and Stalinism together, placing the evolution of the Soviet Union into a totalitarian state at the center of their analysis of the world situation.[103]

This is to say that Huxley was writing in a genre. Unlike other intellectuals of the time who were mounting their critique of Soviet ideology and proclaiming "the end of ideology," he advocated an ideology on his own.[104] Weaving together his views on evolution and the rhetoric of "two camps," however, he advocated an ideology of evolutionary humanism as a counterweight to totalitarian ideologies that indoctrinated and manipulated people. As the head of UNESCO, he launched a project he saw as carrying on this ideology: writing the evolutionary history of humanity.

Huxley's Evolutionary History

Soon after Huxley returned from Eastern Europe in the summer of 1948, his brother, Aldous (who had also attended the Wrocław meeting), planted an idea in his head. Julian, Aldous suggested, should extend his 1946 inaugural address to UNESCO, in which he outlined a vision for the philosophy of a new international organization, into a full-fledged book. Such a book would offer a counternarrative to the Soviet-led peace offensives by promoting UNESCO as the true vehicle to a lasting peace.[105] Julian liked the idea, but the Lysenko book consumed his energies. He decided to unearth the sixty-page-long text he had written for the preparatory commission for UNESCO and publish it as it stood.

Written in the aftermath of his encounter with Lysenko, *UNESCO: Its Purpose and Its Philosophy* (hereafter *UNESCO's Philosophy*) wove together Huxley's views of Darwinian evolution as encompassing both the biological and the social/spiritual realms of the human life with the rhetoric of the two camps.[106] The text laid out what Huxley saw as an ideological counternarrative to totalitarian ideologies. Published in the tumultuous year of 1948, the book resonated with the new anxieties of the early Cold War.

First, consider the evolutionary argument in *UNESCO's Philosophy*. In *Evolution: The Modern Synthesis*, Huxley had sought to reconcile the Darwinian natural selection framework with newer insights from Mendelian and evolutionary genetics. Along with other texts in this genre, starting with Dobzhansky's *Genetics and the Origin of Species* (1937), Huxley's book brought together the findings from a wide range of disciplines—physiology, ecology, systematics, paleontology, cytology, anthropology, and ethology—and related them to a general neo-Darwinian framework, laying out the foundation of the evolutionary synthesis.[107] In contrast to the other biologists who articulated a synthetic framework for biology in the late 1930s and 1940s, however, Huxley portrayed the larger humanistic implications of biological evolution as an intrinsic part of the synthesis. The last chapter of his book, titled "Evolutionary Progress," presented a vision of human progress as the culmination of biological evolution, with Darwinian selection underlying both natural and cultural processes.[108]

In *UNESCO's Philosophy*, Huxley extended his views on the broader cultural implications of the evolutionary synthesis to the case of international organizations and internationalism. UNESCO, he argued, was itself a product of cultural evolution. An innovative type of social organization, conceived as a vehicle to foster an internationalist spirit, and based on ideas of equality grounded in the common evolutionary descent of humankind, it was a driver of the evolutionary progress.[109]

The most consistent element connecting the various strands of *UNESCO's Philosophy* was a very specific critique of Soviet communism as a totalitarian ideology that Huxley described as an impediment to cultural evolution. He decried dialectical materialism as a universalist scientific framework running amok, turning into a conservative dogma. "Dialectical materialism was," he wrote, "the first radical attempt at an evolutionary philosophy." But it was "based too exclusively upon the principles of social as against biological evolution." The fixation on socioeconomic factors turned Marxism-Leninism into a rigid dogma, effectively eliminating diversity and variation in the intellectual sphere. Seeing diversity as indispensable for biological evolution, Huxley argued that the same is true for cultural evolution.[110]

Huxley argued that the new organization should be based on "scientific world humanism, global in extent and evolutionary in background." UNESCO's philosophy was a "synthetic philosophy" with a "truly monistic, unitary philosophical basis." It would ground "the emergence of a single world culture, with its own philosophy and background of ideas," and provide guidance and a philosophy of life for the sociopolitical reality stuck between "conflicting philosophies of East and West." That philosophy would transcend the rivalries of "the American versus the

Russian way of life; or capitalism versus communism; or Christianity versus Marxism." UNESCO's mission was to reconcile "two opposing systems . . . two opposing philosophies of life [that] confront each other from the West and from the East."[111]

UNESCO should, Huxley thought, be concerned not only with the present and the future but also with the past. In *UNESCO's Philosophy*, he stated: "The chief task before the Humanities today would seem to be to help in constructing a history of the development of the human mind, notably of its highest cultural achievements."[112] By removing both nation and race as meaningful historical lenses, such a history would restore a lost sense of common humanity to a world emerging from the rubble of the war.

Huxley's vision for a project of writing a new, evolutionary version of human history was well on its way to becoming reality by the time *UNESCO's Philosophy* was published. At the second UNESCO assembly, held in Mexico City at the end of 1947, the delegates adopted a resolution concerning scientific and cultural history. It outlined the agency's interest in producing a history that would emphasize "the scientific and cultural aspects of mankind" and promote an understanding "of the mutual inter-dependence of people and cultures and of their contributions to the common heritage [of humankind]."[113] After returning from the assembly to UNESCO's headquarters in Paris, Huxley triumphantly announced to an audience at the Sorbonne that "a very considerable sum of money has been voted" by the assembly for studying, and "getting across, the social implications of science." It was to be done, he continued, "in a really big way through press, radio and film, getting the best writers, the most powerful radio broadcasting companies, the best film producers and so on, to collaborate on this important task": "Thus we are starting on something which I regard [as] very important—the draft skeleton for a cultural and scientific history of mankind."[114]

Yet, until the summer of 1948, the project remained largely Huxley's personal obsession. The increasing Cold War tensions and the Lysenko controversy in the Soviet Union gave the project momentum. In the years that followed, the UNESCO scientific and cultural history project started to take shape. It was the first of the agency's high-profile programs dealing with the past, of which the most consequential was the program aimed at identifying and preserving ancient cultural ruins around the globe. Huxley's pet project was also directly tied to another ambitious project: the revision of history textbooks. In 1949, UNESCO called on its member states to reconsider existing practices in history education that put "too much emphasis on military and political factors [which tend to] divide the nations from each other and give too little attention to the history of . . . science, technology, and the arts, for instance—which tends to unite the nations."[115] The content for new history textbooks would be supplied by the scientific and cultural history project. In the new historiography produced by an international team of scholars and scientists, science would replace nations as the great equalizer and unifier of people from different cultures. Writing in early 1949, Huxley emphasized: "Such History must be written *before* really adequate school textbooks can be produced."[116]

If such a history was to be written, it would require a team effort. Huxley thought of it as collaboration on a very large scale. He enlisted, or tried to enlist, large number of scholars to join the project. Ultimately, more than three hundred scientists, historians, and other scholars and education experts from more than fifty countries were involved in one way or another. Never before had a history project attracted so many resources and involved so many participants. The next chapter examines the minutiae of the operations of this rare exemplar of a large-scale collaborative humanities project.

5 The UNESCO "History of Mankind: Cultural and Scientific Development" Project

When in 1948 Julian Huxley proposed producing an evolutionary history of humanity, he set in motion one of UNESCO's major projects of the 1950s and the early 1960s, the "History of Mankind: Cultural and Scientific Development" (hereafter the History of Mankind project). Over the years, the project involved the era's leading historians but also area specialists, philosophers, economists, artists, writers, and cultural ambassadors from Europe, Americas, East and South Asia, Africa, and, in its later stages, the Soviet bloc countries. It was and remains, arguably, the largest international collaborative historical project ever undertaken.

In the mid-twentieth century, the physicist Alvin Weinberg reflected on the dramatic increase in the scale of scientific research since the end of World War II and wondered whether what he called "Big Science" was the most appropriate way to do science.[1] The phrase became a moniker for scientific undertakings that are costly, involve hundreds or even thousands of investigators, and adopt a corporate-style management structure.[2] While many scientists disliked the corporate and collective style of the new way of doing physics, organizing large-scale collaborative efforts to exploit the specific skills of individual scientists and managing large interdisciplinary (and often international) teams played a crucial part in transforming physics into a Big Science and was an important symbol of this mode of working and making knowledge. Significantly, the History of Mankind project involved a form of history writing that was similarly collaborative on a large scale. At a time when physicists were discussing the advent of Big Science, the UNESCO history project was a signal example of what might be called *Big History* in this specific sense. Historians involved in the UNESCO project were experiencing some of the same challenges that drove scientists' anxious reflections on Big Science.

The experience of Big History was a mixed bag, similar to that of Big Science. While the published volumes of the History of Mankind project had little or no historiographic impact, the international, multilingual journals created to manage and organize the collaborative work became spaces in which critical work advancing the critique of Eurocentric perspectives on world history embedded in the project's very design was circulated for the first time. This critique, which later became central for successive generations of world historians, was shaped by the distinct but overlapping contexts of the Cold War and postcolonial nation building and reverberated in the echo-chamber of Cold War internationalism.

History by Committee

In the late 1960s, the Swedish cartoonist Nils Melander depicted the working routine of the History of Mankind project (see fig. 4). In the cartoon, historians representing different cultures are shown in their natural habitat—the library—immersed in reading. Most of them are gathered at the same table, elbows almost touching. They do not interact with one another, however. Those who have spared a moment from their reading to check a book or to take a stroll gaze past each other.

The cartoon can be read as a gentle parody of UNESCO's hallmark feature, cross-cultural collaboration. The organization's premise, expressed in its constitution, was that a major cause of conflict within and between nations was a lack of mutual understanding.[3] In its different programs, UNESCO assembled representatives of different nations, seeing international collaboration as the end as much as the means of its programs.[4] Putting people together in one room is, however, not enough to make them interact productively. While gently mocking UNESCO's naive cross-culturalism, the cartoon inadvertently highlighted the project's most innovative feature: its collective, collaborative nature.

Historians, indeed, rarely work in teams. Or do they? Though the large-scale, international collaborative research projects that came to be associated with Big Science may have felt very new to both scientists and

FIGURE 4 A cartoon by Nils Melander depicting the teamwork involved in the UNESCO History of Mankind project, n.d. (1960s). Image courtesy of the UNESCO Archives.

historians in the years immediately following the end of World War II, this mode of research had a history. As Lorraine Daston has observed, in the first wave of Big Science in the nineteenth century, the human sciences, history in particular, were at the forefront of the trend.[5] The very term *Big Science* was first coined in the last decade of the nineteenth century by none other than the historian Theodor Mommsen, the venerable president of one of the first international gatherings of historians championing scientific history (see chapter 2 above). Addressing the Berlin Academy of Sciences in 1890, Mommsen observed that Big Science (*Grosswissenschaft*), which "cannot be achieved by the lone individual," required institutional backing and state funding.[6] He spoke from experience, as a founder and the director of a monumental endeavor: assembling ancient Latin inscriptions from all corners of the Roman Empire into the comprehensive *Corpus Inscriptionum Latinarum*. A huge, expensive, collective, and long-lived project—indeed, it is still ongoing—the *Corpus* stands as an exemplar of Big Science of its day.[7]

When Huxley originally conceived the History of Mankind project, he might have had another example of a collaborative historical work in mind: H. G. Wells's *The Outline of History* (see chapter 4 above). In the aftermath of World War I, Wells believed that a new version of world history was necessary, one that would not be a patchwork of national (and nationalistic) histories lacking a unity of conception. Weaving together human history, the history of the universe, and the natural history of life on earth, the *Outline* was to provide a common understanding of history and the human condition as a precondition for international peace. It was, of course, one thing for a middle-age political propagandist and writer of scientific romances to conceive such a project and another for him to produce it. As we have seen, however, Wells was willing to delegate.[8] Working on the *Outline*, he enlisted experts from various disciplines. Besides historians, he called on anthropologists, biologists, and practitioners of the new social sciences. Among his principal helpers were the leading classical scholar Gilbert Murray, the historian and anthropologist (and colonial official) Harry Johnson, the zoologist E. Ray Lankester, and the political scientist Ernest Barker. He also consulted dozens of other scholars on specific questions, always paying them handsomely. Most important collaborators received generous credit in the finished product, but Wells retained sole authorship.[9]

Like Wells, Huxley at first envisioned the History of Mankind project as a collective effort resulting in a publication with a single named author—himself. In notes on the project made in May 1948, he wrote: "It seems to me essential . . . that the work should be entrusted to one man, who can of course have recourse to experts and engage helpers." He specifically emphasized: "We do not want a series of volumes by separate authors."[10] He changed tactics in the fall of 1948. In the wake of the dramatic political events of that summer, he moved to scale the project up. His model of collaboration had also changed.

There are parallels—and, indeed, direct overlaps—between Big Science and the organization of the History of Mankind project. When Alvin Weinberg reflected on the effects of the increased scale of scientific research in the United States, he observed that physicists and administrators "have evolved institutional and other devices for coping with broad issues of scientific choice."[11] Weinberg had extensive experience with one such device, the President's Science Advisory Committee (PSAC), whose subcommittees and panels decided, as he put it, "which projects to push, which to kill."[12] That experience informed his critique of what he called "big-scale science's triple diseases—journalitis, moneyitis, administratitis."[13] "The big scientific community tends to acquire more and more bosses," he explained, "[men who] must serve on committees that recommend who should receive support, who should not, [who should] travel to Washington[, etc.]": "In short, the research professor must be an operator [i.e., an administrator] as well as a scientist."[14]

Scholars involved in the History of Mankind project resorted to the same institutional device—committees—to make their choices.[15] What eventually became the International Commission for a History of the Scientific and Cultural Development of Mankind worked for almost two decades (1950–69) and sprawled to include dozens of scholars involved as advisers, corresponding authors, author-editors, and members of the executive committee, as well as administrators and managers overseeing the work. Huxley often mused about using the new managerial techniques and information-processing technologies that would become emblematic of Big Science.[16] His initial plan, however, was decidedly "Little Science."

Huxley assembled the project's first consultative committee in May 1948. The committee, headed by Huxley himself, was small and functioned mainly as a support group for him. It included Joseph Needham, the French physicist Pierre Auger, and the French philosopher and League of Nations veteran (now a member of the UNESCO secretariat) Jean-Jacques Mayoux. The committee stood by Huxley's proposal: a three-volume synthetic work, with one volume providing an overview of the emergence of the modern Western world after 1450, another zooming in on human interactions with the environment, both cultural and physical, and a third presenting a big-picture, comparative history of different civilizations.[17]

In the fall of 1948, in advance of the general UNESCO assembly in Beirut in November, Huxley amassed a larger group to discuss possibly scaling up the project. He asked his confidantes Joseph Needham and his brother Aldous (by then based in the United States) for the names of European scholars who might help in developing a new proposal. By then, Needham, who had been involved in UNESCO since its inception, became his closest ally on the project, sharing his conviction that producing a new history of the world was crucial to UNESCO's mission. First on Needham's list was Lucien Febvre, who, Needham explained, "founded and edited the *Annales d'Histoire Economique et Sociales*, which has paid much more attention than [any historical] journals . . . to the history of technology."[18] Needham, who saw the history of technology as one of the unifying themes of the History of Mankind project, was eager to have Febvre on board. Febvre, who since the end of the war had been leading an effort among French historians to align the *Annales* program with the new politics of "après-guerre," eagerly accepted the invitation. In addition to Febvre, Huxley invited the French historian George Salles, the Swiss diplomat and League of Nations veteran Carl Burckhardt, the Egyptian writer Taha Hussein, and the Mexican archaeologist Pere Bosch-Gimpera to join the planning committee.[19]

But, as Huxley brought in more collaborators, he also relinquished control of the project. The new members of still small planning committee had their own visions of the project's organization and outcome, as did Huxley's friend Needham. Following his work in China during the war as a director of the Sino-British Scientific Cooperation Office, Needham had published a short book, *Chinese Science* (1945), in which he stressed the importance of Chinese scientific achievements and inventions.[20] He now imagined a massive multivolume work on the history of science and technology in China. From Cambridge, he wrote to Huxley in his Paris office: "I . . . would send you a copy of the epitome of the book on which I am now engaged, 'Science and Civilisation in China.' I suggest that this should be typed in a sufficient number of copies for the committee as a concrete example of the kind of basic studies which will be required if the book is to live up to its aims as covering the whole of mankind."[21] Needham proposed that other authors could undertake similar research on other great civilizations. Instead of a three-volume series focusing first on the Western world and then proceeding to a comparative history, he suggested an unspecified number of volumes that would independently trace the development of the world's great civilizations.[22]

When the planning committee met on October 25 in Paris, the consensus was to support the general points of Needham's proposal: that the project would reject a view of dominant and subordinate cultures, that it would give pride of place to previously understudied cultures, and that it would emphasize the unity of all cultures.[23] Finding the plan too general and having no practical plan of its implementation, Febvre raised concerns, but the overall scheme was adopted by majority vote.

Needham, who had by then left his position at UNESCO, was nevertheless entrusted with presenting the project at the general assembly in Beirut in November as the main author of the proposal.

In Beirut, however, the project was not well received. Several delegates—including the representatives of the United States, Australia, Egypt, South Africa, and Norway—expressed doubts that the plan was realistic. The suggestion of the Norwegian delegation was that, instead of producing history texts, the project might focus instead on the more realistic task of compiling and publishing current bibliographies on the proposed themes.[24] Despite his earlier critique of the proposal for precisely that reason, Febvre defended it passionately, argued that the existing scholarship provided a sufficient basis on which to start the project. As he put it: "The progress realized during the last years allows us to write a general history of humanity to show the contributions of all the regions of the world, of all the civilizations or fractions of these regions, to the constitution of the common cultural patrimony. . . . Unceasing exchanges and loans, . . . migration of ideas and instruments, . . . show that nothing exclusively belongs to one single nation."[25] Later, when he reported back to the French national commission on the results of the assembly, he noted, without undue modesty, that it was "only thanks to the French delegation" that the project was saved.[26]

With UNESCO's support for the project assured, Febvre now took the lead in its planning. For help in preparing a detailed proposal, he turned to the ethnologist Paul Rivet, the founder and director of the Musée de l'homme in Paris, with whom he had been collaborating since 1929, when they both participated in Berr's first week of synthesis (see chapter 1 above). Febvre had come to the conclusion that the European and more generally Western civilization had been profoundly delegitimized by World War II.[27] He saw the History of Mankind project as an opportunity to produce a genuinely non-Eurocentric history of the world. In his report to the French national commission to UNESCO, he elaborated his vision of the project: a series of monographs to be written by teams of authors that would "show that, since time immemorial, peoples . . . have communicated by exchanges and borrowings from one another, exchanging tools, technologies, domesticated animals, or improved plant specimens . . . [and] that there are no insignificant people, no poor or destitute civilizations that have not . . . contributed in one way or another to the building of our great and overconfident civilizations that, in fact, survive by borrowing."[28]

Febvre elaborated on the proposal envisioning a grand work of synthesis that centered "the history of peaceful relations" on communication and the exchange of knowledge between virtually all cultures throughout the centuries. The first volume would be methodological, outlining the contributions of different disciplinary methods—biological, anthropological, archaeological, and linguistic—to reconstructing the deep history of humanity. The next two volumes would focus on "everything that circulated from one group to the other"—beliefs, techniques, scientific knowledge, material objects, animals, plants, tools, agricultural and dietetic practices, and so on. "From all that," he explained, "will emerge the image of a humanity in motion from its origins, constantly shifting about in an endless series of transcontinental migrations." Finally, the last volume "would conduct a synthetic regrouping in a *geographic framework*," showcasing what each culture had received from and contributed to the others.[29]

To be sure, the projected history—spanning long time periods, various continents, and many disciplines—would be impossible for any scholar, no matter how erudite or diligent, to produce alone. Febvre stressed that collaboration was the key to the project. "A work of such magnitude," he argued, "could not be prepared in any other way except by an international and collective effort. . . . Only with the help . . . of a worldwide institution like UNESCO could this history of humanity . . . be prepared by genuine scholars who base their [synthesizing] work on original research."[30] The work would be led by a Paris-based international team that would coordinate the overall effort relying on UNESCO's national commissions to keep it informed.

Febvre's proposal, slightly revised by Needham, was distributed to UNESCO's national commissions in the last days of December 1949.[31] The response was a few sympathetic notes and a large number of critical, even hostile reactions. The problem was conceptual as well as organizational. Reviewers pointed out that the supposedly international planning committee was hardly representative of the world cultures. The response of the Pakistani national commission, for example, stated: "The experts on the Planning Committee . . . represent at most the scholarship of Europe, and not even of all Europe, since the Scandinavian countries and Germany did not find a place on the Committee. As for the Middle East, the East and the Far East, they have been completely ignored. Does this mean that in the opinion of Unesco such countries as Arabia, India, Pakistan and China are unworthy of representation?" Allan Kieser of the South African national commission put it more bluntly: "Only a new imperialism of the strongest . . . can give the world [this proposal for] a 'Unitary History' from above." These critiques reflected, at least in part, ongoing debates within UNESCO about decolonization and strategies for incorporating newly independent countries into its structure. Indeed, in the early 1950s, the body's self-conception and mission shifted from its earlier emphasis on "one world" governance to strengthening national identities.[32]

If some UNESCO delegates found the proposal far too conservative, commentators from the United States and the United Kingdom found it far too radical. Ernest Barker, a former consultant for *The Outline of History* who after World War II headed the British national commission, attacked the planning committee for being led by "scholars known for their extreme left persuasions."[33] This was primarily a dig at Needham. As a member of the Cambridge University Communist Group, Needham had been forced to resign from his position at UNESCO after the CIA warned President Truman that the organization was being infiltrated by Communists.[34] Taking his cue from similar allegations of a pro-Communist bias within UNESCO, Barker argued that it would be better to hand this important project over to a body completely independent of UNESCO, perhaps, for example, merging it with the project that he was currently directing.[35] At Barker's direction, the British commission submitted a highly critical evaluation condemning the enterprise as "biased" and "tendentious."[36]

The project continued to evolve over the next few years and was reorganized once again in the aftermath of the general assembly that met in Florence in May 1950. The planning committee was replaced with a new body, the International Commission for a History of the Scientific and Cultural Development of Mankind, which did not include any of the earlier members, with the exception of Huxley, who kept a largely honorary position.[37] This new commission was a corporate body administratively separate from UNESCO. It rented office space at UNESCO headquarters in Paris and remained subject to the general assembly's budgetary oversight, but it now operated as an independent contractor. Needham was given a subsidiary role as a corresponding member. Febvre also lost his place, the grounds cited being his advanced age.[38] The commission was presided over by the Brazilian delegate to UNESCO, the biochemist Paulo E. de Berrêdo Carneiro, with the Yale historian Ralph Turner, a US national representative, appointed as editor in chief. Other new members included the physicist Homi J. Bhabha (the proponent and architect of India's nuclear program who had been suggested by Needham) and the historians Charles Morazé as a representative of France, Constantin K. Zyrayk of Syria, Silvio Zavala of Mexico, and Mario Praz of Italy. Three additional seats were reserved: one for a Chinese scholar, one for a "Marxist from a Slavic country," and a third to be decided after the first two had been filled. To widen the committee's geographic representation, the group resolved to seek more corresponding members from around the world.

The new planning body also reconceptualized the project. Instead of Febvre's emphasis on exchange and circulation, Turner pushed for a chronological organization that would emphasize "the specific contributions of various civilizations" to modern science and culture.[39] Reporting on these developments to Febvre, Morazé, Febvre's right-hand man at the *Annales*, bitterly noted that the new plan was essentially intended to show a linear progression of mankind from its prehistoric state to its "preliminary" climax—"the American way of life."[40] Turner, in turn, bitterly commented on the lack of cooperation on the part of Morazé and Febvre.[41]

While losing the battle over the grand design of the project, Febvre did, however, receive a compensation of sorts: a new scholarly journal funded by UNESCO that was to serve as a testing ground for the project. As the head of the journal, Febvre was in his element, and he used the journal to implement his vision by other means.

Febvre's Cahiers*: Historical Journals and the Making of Historical Knowledge*

Febvre had strong opinions about how a collective historical work should be organized. Unlike Huxley, he believed that the project should involve the largest possible number of scholars from around the world as true collaborators at every step of the way, from collecting and accumulating the information to processing and interpreting it. Like Huxley, he proceeded from his previous experiences of collaboration: the work he had been orchestrating through the *Annales* since the journal's inception. When he and Marc Bloch founded the *Annales* in 1929 with the ambition of altering the intellectual terrain of history by encouraging cross-disciplinary communication, they introduced specific practices of historical research as well. One of the first projects of the *Annales* was "the study of the present" ("une enquête contemporaine"), focused on the impact of the Great Depression in different parts of the world and its ongoing effects. Through the journal, Bloch and Febvre solicited contributions from amateurs, professional historians, and experts in various topics.[42] It was in this spirit that Febvre solicited an article from André-Georges Haudricourt, an agronomist-turned-ethnobotanist who had spent a year in Leningrad studying Vavilov's methods for a collective survey of the agricultural techniques used throughout the world (see chapter 3 above).[43]

Teamwork became the hallmark of the historiographic school that formed around the journal. The collaborations quickly grew international. In one of the first issues of the *Annales*, Febvre put forward Latin America as a "privileged site of study," a vantage point from which to engage in cross-cultural comparisons.[44] As he did with other collaborative projects orchestrated through *Annales*, throughout the 1930s and 1940s he regularly commissioned reviews and articles on Latin America.[45] Most reviews were written by French academics who had taken jobs abroad to escape economic depression and political instability at home. In 1935, Fernand Braudel and Claude Lévi-Strauss took professorships in history and sociology, respectively, at the University of São Paulo in Brazil. Prior to that move, Braudel had spent a decade in Algeria, teaching at a lycée in Constantine. His transnational trajectory had a crucial impact on his work.[46]

Febvre brought his editorial approach to gathering information and organizing collective projects to the History of Mankind project. As part of the reorganization in 1950, and in order "to give Scientific History a base," the new commission established its own journal, *Cahiers d'histoire mondiale/Journal of World History/Cuadernos de historia mundial*, with Febvre as editor in chief.[47] While the journal's format was unusual—it accepted articles in French, English, or Spanish with abstracts appearing in French, English, Spanish, German, Russian, and Arabic—it was intended, much like the *Annales*, as a place to publish working material commissioned by the project leaders, in this case, the author-editors of the History of Mankind volumes and the international commission coordinating the project. Most requests concerned specific regions about which there was little or no historical research available in French, German, or English, the main languages of the world historical community at the time. One of the first issues of the *Cahiers* featured an extensive review (in French) of literature (in Russian) on the ancient civilizations in Central Asia, written (originally in German) by the Austrian ethnologist Karl Jettmar, fulfilling the request of the British archaeologist Jacquetta Hawkes, one of the three author-editors of the volume focusing on ancient human civilizations around the world.[48] Neither Hawkes, who specialized in Neolithic European history, nor her coauthors, the Dutch Egyptologist Henri Frankfort and the British archaeologist Leonard Wooley, who specialized in ancient Mesopotamian civilization, had expert knowledge of the region or read Russian.

Even while the division of labor kept Febvre at bay, the frictions between him and the international commission were very much apparent.

In a lengthy editorial in the first issue of the *Cahiers*, he ruminated about his role being "paradoxical," given that, as the editor in chief, he had relatively little control over the publication's content. The research articles and review essays were commissioned by the author-editors or by members of the commission. That is to say, he lamented, the articles were being chosen "without his participation, and from people he usually does not know." The editor "can, to be sure, exercise his criticism; if [the articles] are too long, he can cut and shorten them, he can, in certain extreme cases, communicate his doubts to the members of the Editorial Board or to the Author-editor responsible for the order, or even, as a last recourse, to the President of the Commission. But this is all." He compared his role to that of "a chef who is required to prepare the meals, four great meals a year, as delicate, and well composed, and at the same time as economical as possible, without having the say as to the buying of the food, nor even as to the menu."[49]

In the years that followed, Febvre struggled to reconcile the *Cahiers*' different functions: as an outsourcing device for capturing material for the History of Mankind project, as its "testing bench" ("banc d'essay"), and, at the same time, as a publicity arm for the project. In 1955, having turned seventy-seven, he distanced himself from the journal and handed it over to Morazé. The incessant fights with other members of the history project's international commission had exhausted him.[50]

Febvre's tenure as editor in chief may have been short, but it made a critical imprint on the project.[51] While his vision for the project was sidelined, he effectively turned the *Cahiers* into a space to explore alternative narrative plots. For starters, he reprinted his own rejected proposal in one of the first issues.[52] From its first issue, the *Cahiers* was divided into different sections with different purposes. Two sections—one publishing original research articles and another commissioned reviews—were beyond his control. But he added two more, "Critiques and Suggestions" and "Official Texts." "Official Texts" provided a space in which he brought to readers' attention updates and information on the status of the History of Mankind project, including minutes, white papers, and other preparatory material. "Critiques and Suggestions," on the other hand, played a role similar to the *Annales*' section "Debates and Combats," which has been one of the successful features of the periodical, encouraging debate in the tradition of combativeness so characteristic of the *Annales* school.[53] The "Critiques and Suggestions" section effectively turned the journal into a discussion forum that Febvre used to initiate new lines of inquiry. He would publish his lengthy editorials there, and it was in that section that he placed his rejected proposal setting up a dissenting tone of the journal from the beginning. The gambit to circumvent the planning committee might have failed had Febvre not succeeded in developing the *Cahiers* into a venue in which questions of the methodology and theory of world history would be discussed in the 1950s and 1960s.[54] The very first volume of the *Cahiers* featured an essay that advocated a novel approach to world history, one that would dispense with the categories *Europe* or *Asia* altogether.[55]

When the essay was published in 1954, the author, Marshall Hodgson, was a young postdoctoral researcher in the Ford Foundation's Comparison of Cultures program at the University of Chicago, working as an assistant to one of the History of Mankind's author-editors, the Chicago history professor Louis R. Gottschalk. Hodgson had trained in Oriental studies at Chicago, the institutional home of the great Egyptologist James Henri Breasted.[56] As a doctoral student in the history of Islam, he developed a distaste for the field, which he found profoundly Eurocentric, shaped by a master narrative that presented modernity as a Western phenomenon. To combat what he called "Western provincialism," he started to work on a book. In 1945, he published an outline of the book, along with a series of accompanying essays, in a small-circulation Quaker teachers' journal.[57] The book would advocate for a different approach to world history, one that would abandon the Mercator projection, reject the division of the world into East (Asia) and West (Europe and the United States), and show instead the continued vibrancy of different world cultures. "THERE IS NO ORIENT," he wrote in the published outline.[58] In the words of the historian Edmund Burke, who rediscovered and published Hodgson's little-known works in the 1990s: "The major purpose of Hodgson's interregional approach was to resituate modernity and to unhook it from Western exceptionalism . . . making it possible to rehistoricize the premodern period as something more than an antechamber to the rise of the West."[59]

Having developed this critique, Hodgson found the experience of working on the History of Mankind project frustrating. The volume assigned to Gottschalk, his supervisor, was a synthetic account of the period from 1300 to 1775 that was described in the prospectus as "treat[ing] the foundations of modern European culture, its development, and its expansion and its early impact on Asian, African and American peoples."[60] Ralph Turner, the project's overall editor in chief, had recruited Gottschalk because of his international reputation as a historian of the French Revolution. Gottschalk, who had no expertise in the non-European world, commissioned a number of articles for the *Cahiers*.[61] Hodgson found Gottschalk's approach deeply problematic. The specialists whom Gottschalk approached with requests were historians or area studies specialists either working or trained in the United States or the United Kingdom. In their accounts, non-European histories entered the narrative mainly as past and dead glories. When Hodgson pointed out to Gottschalk that the material commissioned for the volume omitted the history of Islam and the role of Islamic cultures in world history altogether, Gottschalk replied that such a history could not be written because of the lack of sources and that, if it were written, no one would be interested in reading it.[62]

The experience of working on the History of Mankind project confirmed Hodgson's worst perceptions of Western provincialism. He reworked one of his essays into a programmatic article titled "The Hemispheric Interregional History as an Approach to World History." Historians, he argued, must reconsider their entire approach to world history. Instead of the loaded and misleading categories *West* and *East*, he called for integrated studies of entire hemispheres. "Historical life," he wrote, "from early times till at least two or three centuries ago, was continuous across the Afro-Eurasian zone of civilization; that zone was ultimately indivisible."[63] Therefore, he argued, historians should conceive of Afro-Eurasian history as an interconnected and interactive entity and study it empirically as a whole.

Hodgson first submitted the paper to the *South Atlantic Quarterly*, only to receive a swift rejection. He was hesitant to send it to the *Cahiers*, anticipating another rejection given that the paper implicitly criticized the History of Mankind project, with which the journal was involved, but decided to submit anyway, scribbling on his copy of the first issue of the *Cahiers*: "MUST PERSUADE these people of the need for perspective." To his surprise, the paper was accepted as quickly as it had been rejected by the *South Atlantic Quarterly*. In a letter relaying the good news, François Crouzet, then a young assistant to Febvre at the Sorbonne and the journal's editorial assistant, wrote that Febvre "much appreciated" the paper.[64] Indeed, Hodgson's critique was quite congenial with Febvre's own vision of the project.[65]

Hodgson's counternarrative of world history anticipated Edward Said's critique of Orientalism by more than two decades.[66] That critique, which formed an important background for the subsequent rise of postcolonial studies, also provided a context for historians' rediscovery of Hodgson's work in the 1990s. Before it was recirculated by postcolonial scholars, however, Hodgson's critique of Eurocentrism was circulated through the international apparatus created for the History of Mankind project, not only in the *Cahiers*, but also in the Soviet journal associated with the project, *Vestnik istorii mirovoi kul'tury*, which promoted its own political and scholarly agenda.

Cold War Internationalism and the Writing of History

During the first years of its existence, the international commission for History of Mankind project did not have any Soviet representatives because the Soviet Union was not a member of UNESCO until after Stalin's death. In April 1954, in the climate of Khrushchev's political "thaw," the Soviet Union officially joined UNESCO.[67] In early January 1955, the commission made a cautious first move when its secretary, Guy Métraux, approached the academic secretary of the Soviet Academy of Sciences, M. N. Tikhomirov, asking for a few articles by Soviet scholars to fill in some gaps in the already-written drafts of the first volumes of the History of Mankind. The author-editors requested specific information: on Byzantine influences on Russian culture, on the history of Mongols in Russia and Central Asia, and on the history of Kievan Rus' (the first organized medieval state located on the lands of modern Russia, Ukraine, and Belarus).[68] The matter was brought before the Presidium of the academy, which found it "desirable" not only to arrange for the requested articles but also to insist on Soviet participation across the board. The academy recommended that the Soviet Ministry of Foreign Affairs, which at the time coordinated all UNESCO-related matters, initiate talks about Soviet involvement. In Paris, Soviet diplomats promptly met with Carneiro, the chair of the international commission, to relay Soviet interest.[69]

The History of Mankind project resonated with several simultaneous developments in the Soviet Union. In the new post-Stalinist climate, politicians stressed the need for Soviet scholars to reestablish contact with "progressive Western scholars" and to project a positive image of the Soviet Union "dismantl[ing] the myth of the 'iron curtain.'"[70] Participation in the project thus presented an opportunity for cultural diplomacy. At the same time, collaborative, large-scale projects were a hallmark of Soviet historiography. As we have seen, since the 1920s, large-scale, coordinated interdisciplinary research initiatives had been prioritized as a distinct feature of Marxist and distinctly *Soviet* historical scholarship. The emphasis on nationwide, coordinated collective research received a major boost from the expansion of history programs in the universities after World War II, which produced large numbers of historians who filled research positions in various projects.[71]

A signal example of Soviet collective interdisciplinary scholarship was the *Great Soviet Encyclopedia* (*Bol'shaia Sovetskaia entsyclodediia*, or *BSE*), a compendium of sixty-five volumes published between 1926 and 1947. Modeled on the French *Encyclopédie*, the *BSE* was launched to promote the values of the Soviet "Enlightenment."[72] In 1949, the second edition was launched in order "to elucidate widely the world-historical victories of socialism in our country . . . in the provinces of economics, science, culture, and art." The president of the Academy of the Sciences, Sergei Vavilov (the brother of Nikolai Vavilov), was the *BSE*'s editor in chief.[73] After Vavilov's death in 1951, another physicist, B. A. Vvedenskii, took over. When negotiations regarding Soviet participation in the History of Mankind project began, the Soviet representatives put forward the *BSE* as an obvious model for their contribution. Carneiro sent the outlines of the projected volumes to the Soviet Academy of Sciences and invited it to select a senior scholar to join the international commission (filling the seat reserved for "a Marxist from a Slavic country"). The academy promptly nominated the deputy editor of the *BSE*, the historian of technology Anatolii Zvorykin (who spoke French fluently).

Anatolii Alekseevitch Zvorykin had a colorful life.[74] He was born in Murom in 1901 and joined the Bolsheviks during the Revolution. During the Civil War, he served in the Red Army as a political worker (*politrabotnik*) while attending classes in one of the *rabfaki*, the postsecondary educational facilities for working-class people created after the Revolution. After the Civil War, he studied economics and the history of technology at the Institute of Red Professors and taught at the Communist Academy in Moscow while working on a dissertation on the socialist reconstruction of the coal industry.[75] In 1934, soon after graduating from the Institute of Red Professors, he was appointed a deputy director of the Institute for the History of Science and Technology in Leningrad. The institute had been created by Bukharin as a successor to the KIZ soon after his return from the 1931 congress in London (see chapter 2 above).[76] It employed many former KIZ members along with new cadres from the Communist Academy.[77] In the spring of 1936, amid Stalin's campaign against Bukharin, the Politburo dismissed the entire Leningrad staff and moved the institute to Moscow, with Zvorykin effectively becoming the acting director. Bukharin rightly regarded this development as marking the end of the institute and complained bitterly to Gleb Krzhizhanovsky and Nikolai Gorbunov about his "inheritance [being] handed over to Zvorykin."[78] Zvorykin duly attacked Bukharin, publishing a statement claiming that, under his leadership, the institute had become an "outpost of fascism."[79] Yet this did not keep him out of trouble, and in 1937, in the midst of the Great Purges, he was fired and expelled from the Party.[80] Still, he was not arrested, and in January 1938 he even got his job back. Two months later, after the Moscow trials ended in a death sentence for Bukharin and other condemned Bolsheviks, the institute was disbanded.

Many militant Marxists of the 1930s who managed to survive the Great Terror did not easily fit into the academic structures of late Stalinism, with its increasing emphasis on technical expertise and specialization. Zvorykin, however, reinvented himself successfully. After a short period of unemployment, he started to teach at the Moscow Mining Institute. At the outbreak of World War II, he enlisted in the army and fought in Stalingrad and on the Central Front. After being seriously injured, he spent the rest of the war working as a midlevel administrator in the coal industry, proving himself an able organizer. And his service was rewarded: his membership in the Party was restored. In 1946, he received a prestigious appointment as one of the editors of the science and education section of *Pravda*. The next year, his spectacular comeback continued with his appointment as deputy editor of the *BSE*.

By the time Zvorykin joined the international team of the History of Mankind project, the project's author-editors had already gathered most of the material for the planned six volumes and were busy writing up their synthetic accounts. When the committee met in Paris, he insisted that the author-editors should incorporate material from Soviet scholars in their drafts and take into account, wherever possible, their diverging perspectives. Back in Moscow, he reported to P. M. Pospelov, a high-ranking party official who was overseeing Soviet work on the project, that the "balance of power" on the planning committee was unfavorable to the Soviet Union since "the Americans and British are playing the leading roles" and the project was already well under way. Yet the situation was not without promise, given that the French had a strong voice on the committee and Turner, the American editor in chief, was, according to Zvorykin, a "progressive scientist" with a record of consistently voting for the Democratic Party.[81] Pospelov gave Zvorykin the green light to proceed.[82]

Zvorykin proved himself a skilled organizer and a good team player. Less than a year after he joined the team, Carneiro recommended that he be made the commission vice president to compensate for the late arrival of the Soviet team. Huxley and Turner objected. While agreeing that Zvorykin had proved to be "a highly efficient addition to the workforce," Turner worried that promoting a Soviet representative to such a powerful position would endanger the entire project since, in his view, the Soviets would inevitably try to impose an alternative philosophy of history on the project. Carneiro nevertheless pushed for Zvorykin, arguing that the commission should cooperate with whoever was willing to participate to make the project a truly international undertaking. With two opposing votes (Turner's and Huxley's), Zvorykin was elected vice president.[83]

Zvorykin recruited dozens of Soviet specialists from diverse fields and institutions as contributors. In addition to historians, he commissioned specialists from institutions ranging from the Institute of Oriental Studies, the Institute of Chinese Studies, and the Ethnographic Institute to the Institute of Complex Problems in Transportation, the Institute of World Economy and Foreign Relations, the Institute of World Literature, the Semashko Institute of Public Health, and the Institute of Experimental Design.[84] These experts, along with various specialists from the ministries and scholarly journals, reviewed drafts of chapters provided by the volumes' author-editors.

By June 1956, Zvorykin had assembled several hundred pages of comments, suggestions, and alternative outlines in response to drafts of the first four volumes of the projected six. Commissioning articles for the *Cahiers*, however, proved more difficult. In May 1956, Zvorykin reported to his Party bosses that many Soviet academics were reluctant to publish in the *Cahiers*. As a typical reason for their hesitation (if not outright refusal), he cited a remark he often heard from Soviet scholars: "Why would I ruin my publication list by publishing in a bourgeois journal?"[85] He turned the problem into an opportunity, however, using that reluctance to publish abroad as justification for creating a Soviet counterpart of the *Cahiers*. The journal, *Vestnik istorii mirovoi kul'tury* (The courier of the history of world culture), published by the *BSE* with Zvorykin as its editor in chief, was launched in January 1957. For Soviet readers, *Vestnik* was a remarkable and highly atypical journal, regularly publishing Western scholarship in the humanities usually not readily available in the Soviet Union. Between 1957 and 1961, in six issues per year, it featured, besides contributions from Soviet scholars, material related to the work of the History of Mankind project: discussion summaries, synopses, drafts, and articles translated from the *Cahiers*.[86]

From the start, Zvorykin intended *Vestnik* to have a political agenda. As he put it in his report to his Party bosses in May 1956, the journal's "main task is to make the contributions of not only the Soviet Union but also China, other people's democracies, and Africa and Asia more broadly represented" in the History of Mankind project.[87] In October, he reiterated: "Our success will depend not only on the publication of our own material but even more so on the promotion of the national groups, first of all those of the socialist countries and the East."[88] In the inaugural issue, which appeared in January 1957, he explained that the goal of the journal was "to publish articles and material on topics and regions that have not been studied sufficiently or at all," it being "necessary to launch intensive studies regarding the history and the cultures of the peoples of America, Africa, the Middle East, and the Far East" who had been understudied or misrepresented in the existing historiography.[89]

Fittingly, the first issue of *Vestnik* opened with a Russian translation of Hodgson's article.[90] In an accompanying editorial, Zvorykin endorsed Hodgson's critique of Eurocentrism: "There can be no such juxtaposition between the people as some belonging to the 'West' and others to the 'East,' some to 'small nations' and others to 'big' ones."[91] Hodgson's article was flanked by Zvorykin's editorial, printed in both Russian and French, and a response to Hodgson by the historian E. M. Zhukov, a specialist on Japan. Zhukov had been the chief editor of the multivolume collective project *World History*, which began appearing in print in 1955.[92] Endorsing Hodgson's critique of the Eurocentric approach to world history, he stressed, however, that Marxist scholars had long argued along similar lines. His own *World History* project, he stated, aimed at "freeing historians from Eurocentrism, the dangers of which were described in such detail by Hodgson."[93]

Over the course of the journal's five-year existence, it kept the emphasis on interconnections across political borders and geographic regions. A sizable portion of the original contributions centered on Asia—for most part Soviet Central Asia but also China, India, Japan, South and Southeast Asia, and Africa. The articles were accompanied by abstracts in either French, English, or German, and each issue additionally included lengthy reviews of Russian-language scholarship in one of those languages. Reviewing the journal in 1958, the British historian Robert Browning noted: "The Soviet contribution to these discussions [of the History of Mankind project] is marked . . . by [Soviet scholars'] insistence on abandoning . . . the old Europe-centered view of history."[94]

To be sure, Zvorykin's strategy of attaining more visibility for the Soviet contribution to the History of Mankind project by centering *Vestnik* on the non-West was politically opportunistic. In the mid-1950s, the ideals of cultural integration and cultural unity had been rekindled in the Soviet Union as part of Khrushchev's strategy of de-Stalinization, which emphasized both the Soviets' alliance with the decolonization movement and the rejection of Stalinist nationalistic, Russia-centered cultural politics. The partnership with postcolonial India in the 1950s stimulated Khrushchev's vision of Asia as a place where the Soviet Union could present itself as an anticolonial power that supported anticolonial movements in the

Afro-Asian nonaligned world.[95] In this political climate, a projection of Soviet-Asian "transnational fraternity" into the past was very welcome.[96]

While in the 1950s studies of the cultural integration between West and East throughout history met the political demands of the day, Soviet historians who contributed to *Vestnik* revived methodological approaches that had deeper roots in Russian historiography. As Vera Tolz has pointed out, from the last decade of the nineteenth century and into the 1930s, members of the Russian school of Oriental studies "wrote extensively about the origins of various definitions of Europe and Asia and, in the process, questioned and rejected the East-West dichotomy as a figment of the European imagination."[97] In the 1950s, these arguments resurfaced in the context of the Cold War and decolonization, stimulating the expansion of the institutional base of Soviet Oriental studies. Sensing an opportunity, Zvorykin enlisted B. A. Vvedenskii, the chief editor of the *BSE*, to lobby for the creation of an "Institute of World Cultures" that would "train specialists who could represent the Soviet Union in discussions about humanism, decolonization, West and East 'civilizations,' and so on."[98]

While the plans for an institute of world cultures fell through, Zvorykin's efforts to press for the Soviet conceptualization of world history to be taken into account by the leaders of the History of Mankind project attained some success. An opportunity presented itself when Gottschalk's volume, which incited Hodgson to publish his critique of Eurocentric history, reached the publication stage. On receiving Turner's invitation in 1950, Gottschalk accepted only "after some hesitation," anticipating difficulties.[99] According to the original schedule, in five years' time he was expected to produce the complete draft to be sent to international experts and circulated among the national committees of UNESCO member states for feedback. Soon after he began receiving the articles on specific topics that he had commissioned for the *Cahiers*, he recognized the problem Hodgson had already identified: most of the contributions were written from a Eurocentric perspective. Writing to Métraux in 1953, he explained: "We know so little about what the world looked like to non-Europeans before the great discoveries. What did the Chinese, Japanese, Arabs, Aztecs, etc. know of each other, of the Pacific islands, of the West, etc. before 1500?"[100] He did not receive an answer to that question but nonetheless decided to proceed with the project, recruiting doctoral students and research assistants at the University of Chicago to conduct original research. As a stopgap solution, they were invited to participate not as research assistants but as authors of the portions of the draft.[101] After several delays, in the spring of 1958 Gottschalk finally produced the draft.

Despite Gottschalk's attempt to counterbalance a Eurocentric orientation by bringing in and motivating younger scholars to undertake original research, many reviewers were dismayed by the submitted draft. The eminent Indian historian Ramesh Chandra Majumdar found it hopeless. The parts on South Asia, he wrote in late July 1958, were reminiscent of "the propaganda leaflets written by the Christian missionaries in India about a century ago." The draft, he felt, could not be improved and should be rejected.[102] Other reviewers tendered a wide range of objections regarding the volume's representation of the non-European past.[103] In fact, Majumdar had written Turner directly, advising: "Scrap the [drafts] altogether and get them written by one who has some knowledge of the subject [as far as] . . . the countries in Asia [are concerned]."[104]

Turner took the criticisms seriously and wrote to Carneiro suggesting that the manuscript be rejected: "We must find a way to get vol. IV rewritten or rather, I think we should say, written, and in a short period of time. These are difficulties no one likes to face, but face them we must." Under the circumstances, he suggested handing the volume over to Zvorykin. His reasoning was twofold. First, he trusted that "the Russians will jump at the chance to write this volume." He continued: "We can take care of other points of view [through] editing." Second, he thought Zvorykin "[could] command scholarly assistance quickly and in an amount that will expedite the preparation for the volume." He estimated that Zvorykin "[would be able] to deliver the manuscript by July 1, 1961."[105] Without further ado, he approached Zvorykin, explaining the situation, and offering full control over the volume in case he took on the assignment. He was hoping the Soviet team could produce a new draft within a year.[106]

Turner's request might be read as suggesting that Zvorykin's effort to use the History of Mankind project to court postcolonial scholars in India and other nonaligned countries had been moderately successful. But, despite having achieved this recognition, Zvorykin now failed to gain the support of his Party bosses, who advised that he turned the offer down. By then, UNESCO itself was under attack in the Soviet Union. Turner's offer to Zvorykin came in the summer 1958. In the fall, the chairman of the Soviet national commission to UNESCO, G. A. Zhukov (no relation to the E. M. Zhukov mentioned earlier), publicly lambasted UNESCO for catering to American interests, sparking a discussion on whether the Soviet Union should leave UNESCO altogether.[107] Behind the scenes, Zhukov argued that the $2 million annual Soviet contribution to the UNESCO budget would better serve Soviet interests if it were diverted to bilateral initiatives involving Asian and African countries, bypassing UNESCO entirely.[108] In 1961, *Vestnik* was shut down. In the end, Gottschalk's draft was published in a revised version.

By the early 1960s, the History of Mankind project as a whole was facing a crisis. As drafts began circulating around the world and readers' reports began flowing back to UNESCO headquarters and to the offices of the author-editors, the project's international commission faced difficulties in trying to represent multiple perspectives while maintaining coherence. When, in 1963, the volumes finally began appearing, the reception was disappointing. Reviewers found the voluminous footnotes that the author-editors used to incorporate diverging views of numerous contributors particularly distracting.[109] In an especially harsh review, the Cambridge historian John H. Plumb ridiculed the project as "an encyclopedia gone berserk, or re-sorted by a deficient computer."[110]

While the published volumes of the History of Mankind project might have left little or no historiographic impact, the journals created in conjunction with the project provided a forum for an important methodological and theoretical discussion in world history. They were perhaps the most successful outcome of the project. Yet they were only one of many other experiments with organizing and managing the labor of vast numbers of researchers across international borders who were involved in the History of Mankind project one way or another.[111]

In his cartoon, Melander depicted the project team in a library, the traditional site of humanist learning. And of course a library epitomizes the humanistic ideal of scholarship as a solitary act of contemplation, the work anything but mechanical. What Melander missed, however, was the project's radical mixing of traditional forms of history writing and humanist learning with collective and corporate Big Science and its division of labor, settings, and technologies. In his presentations, Huxley liked to muse about the accelerated mechanization and information processing characteristic of the age. As the director of UNESCO, he envisioned that the organization would promote the spread of new mechanical technologies for information processing to disciplines beyond physics—first of all biology.[112] The reference to a computer in Plumb's review was hardly accidental. Electronic computers and electromechanical devices for processing increased amounts of information were important features of Big Science. The next chapter shows that the new data- and information-management technologies became entangled with Big History as well.

6 Information Socialism, Historical Informatics, and the Markets

By the time the first volumes of the History of Mankind began appearing, scientific history was back in vogue, this time as a data science. The proponents of the new economic history, econometrics, and cliometrics extolled quantification—the translation of arguments about historical phenomena into statements about variables and quantifiable units of analysis. This new trend, as one reviewer wrote in 1966, was characterized by attempts "to state precisely the questions subject to examination and to define operationally the relevant variables," "to build explicit models that are relevant to questions at hand," and "to test the models . . . against the evidence."[1]

Quantitative history has a long history.[2] After World War II, new technologies for organizing, managing, storing, and processing data—photocopying machines, microfilm-based devices, and, above all, electronic computers—promoted quantitative analytic procedures that had developed earlier and independently. New technologies of quantification, datafication, and information management enhanced already-existing quantitative conventions in history, but the relations between informatics and historiography were neither simple nor technology driven. This chapter examines some of the entanglements between history writing, computing, and politics by following a thread that takes us back one more time to the 1931 Second International Congress of the History of Science and Technology.

Bernal's Information Socialism: From London 1931 to Cold War America, via Russia

For the British crystallographer and polymath John Desmond Bernal, the 1931 London congress was a life-changing event. Unlike other British participants discussed earlier in this book, Bernal was a committed Communist who had joined the Communist Party of Great Britain as a freshly minted Cambridge graduate in the early 1920s.[3] In his report on the congress, he characterized the event as "the most important meeting of ideas that has occurred since the Revolution."[4] After the congress, he went to the Soviet Union, joining the second of the two group tours organized by the science journalist James Crowther.[5] In August 1931, following in Julian Huxley's footsteps, he visited some of the same scientific institutes and laboratories in Moscow and Leningrad. He traveled to the Soviet Union again the following year and made many such trips throughout his life.[6]

Like Huxley, Bernal was attracted to the Bolsheviks' dream of a scientific socialist society. Unlike Huxley, who described the realities of Soviet society as "terribly depressing," he found the Soviet Union "grim but great."[7] In 1932, soon after his second visit, he began to work on a book critiquing the exploitative practices of science under capitalism and elaborating on how, under socialism, science could be harnessed for the benefit of society. That book, *The Social Function of Science* (1939), also launched his lifelong campaign for reform of the traditional ways in which scientists communicate and share information among themselves and with others.[8]

Bernal believed that the principles of socialist planning, distribution, and ownership should be applied to the sphere of information. *The Social Function of Science* included a chapter devoted to scientific communication. In it, Bernal argued that the increased number of specialized disciplines, the multiplicity of languages, and the scale and scope of scientific publications made traditional forms in which scientific information was distributed (the scientific paper and the scientific journal) "increasingly cumbersome and wasteful." "Capitalist practices" of managing scientific information, such as secretive government operations, bureaucratic rivalries, patents, and any form of intellectual property protection (including copyright), he held, impeded the free flow of scientific information.

Condemning these practices as symptoms of the crisis of capitalism, he advocated an alternative, socialist approach to scientific information. It would be based on the principles of centralized planning and equal access. Instead of scientific journals, a network of information offices coordinated from a center would handle the submission, selection, and distribution of "publication units." Instead of scientific papers—the coherent accounts of problems and the results of finished work—the legitimate and desirable form of publication would be a brief note on ongoing research or preliminary research results. Bernal envisioned a worldwide "newspaper" that would communicate scientific results efficiently and without delay, making scientific knowledge available to everyone.[9]

Bernal's vision of "information socialism" built on a variety of contemporary visions and actual programs. In *The Social Function of Science*, Bernal acknowledged the influence of Gordon Watson Davis, the director of the US Science Service. In the 1930s, Davis formed the nonprofit Documentation Institute as well as the Auxiliary Publication and Biblioform Services, which began using microfilm technology to copy and disseminate scientific publications. In 1933, he circulated "A Project for Scientific Publication and Bibliography," which included a proposal for creating one central organization (which he called the "Scientific Information Institute") to distribute scientific information nationwide and worldwide "in the most effective and efficient manner." Bernal reprinted Davis's proposal as an appendix to his book.[10] The proposal called for, in particular, the adoption of microfilm technology on a large scale as a modern solution to the problem of handling the increasing volume of scientific publications.[11] In his book, Bernal endorsed the idea of using microphotography and filmography to facilitate and expedite the flow and exchange of scientific news. Combined with "modern business filing methods" and "automatic machinery," the mechanization and automation of the management of scientific information could save scholars "months and years of scholarly research," he reasoned.[12]

In the 1930s, many scientists and administrators thought that the days of the individual researcher had passed. In the future, science would have to be organized more like a large industry, with information accessed, distributed, and managed accordingly.[13] A number of progressives and socialist activists of various backgrounds on both sides of the Atlantic imagined and some attempted to build a global information network as a prerequisite for the realization of the internationalist dream of a single world government.[14] The interwar era's mingling of the informationist and the socialist visions was captured by H. G. Wells (a Fabian socialist) in his collection of essays *World Brain*, in which he argued that a unified system of information was a critical step toward world peace and the establishment of a unified world government.[15]

In Britain, Bernal and his proposals were popular among young documentation specialists and librarians.[16] The discussion of his ideas in these circles was interrupted by World War II, which broke out soon after the publication of *The Social Function of Science*. Toward the end of the war, however, Bernal's ideas gathered momentum in the context of the discussions that spread from within the British Council. The question of information was put on the agenda of the council by none other than Bernal's guide in the Soviet Union, the science journalist Crowther, who at the beginning of the war was appointed secretary of the council's newly founded Science Committee (apparently his employers did not know about his membership in the Communist Party, which he had joined in the 1920s).[17] From this position, he launched an all-out campaign for a centralized national system to facilitate access to scientific information in Britain, hailing the organization of scientific information as "not merely a matter of convenience for a few experts" but "the key to the future of humanity." Addressing the conference of the British Association of Special Libraries and Information Bureau (ASLIB) in 1943, he called on "bibliographers and scientific informationists" to get "their proposals on to the agenda for the forthcoming settlements of the world, and to have their proposals ready."[18]

In the optimistic climate of postwar Britain, when the adoption of job security and comprehensive welfare policies seemed to be the nation's most likely future, Bernal's proposals for information socialism found receptive audiences. Bernal was the keynote speaker at the first postwar ASLIB conference in June 1945. He used the opportunity to revisit proposals for information management he had first elaborated in *The Social Functions of Science*. He continued to publicize his ideas about a centralized national information system within the Association of Scientific Workers, which organized two conferences—in October 1947 and April 1948—focused on problems of science communication. He spoke at both meetings, elaborating proposals for a centralized British science information service.[19] By this time, even the Royal Society, that stronghold of traditional British scientific institutions, engaged in debates about the potential for a national center, what Bernal called an *institute of scientific information*.[20] When the society began planning a conference on scientific information in the fall of 1947, Bernal was on the organizing committee.[21]

The planning of that conference, scheduled for the summer of 1948, coincided with increasing tensions between the Soviet Union and Britain.

In the tense political climate, and amid the deteriorating relations between wartime allies, members of the organizing committee received Bernal's circulated paper[22] with "grave disquiet." "What a planner's paradise is here envisioned," wrote F. M. R. Walshe, lambasting, in particular, Bernal's suggestion to "replace the publication function of the scientific journal" with a "national distribution authority for scientific information" that would be responsible for the "production and distribution of the basic unit of scientific information—the individual paper." Walshe warned: "There are possibilities of censorship . . . [, which] is far too authoritarian. . . . The possibility of a central body deciding whether or not a paper should be published is surely abhorrent."[23]

The condemnation of Bernal's proposals as antidemocratic and authoritarian originated not with Walshe but with the philosophical chemist Michael Polanyi and the biologist John R. Baker, who started a campaign against Bernal shortly after the publication of *The Social Function of Science*.[24] Polanyi and Baker tirelessly organized against not only Bernal but also the larger threat of "Bernalism," forming the Society for Freedom in Science (SFS) for that purpose.[25] Baker described Bernalism as "the movement against pure science and against freedom in science [that] was first brought to Great Britain by the Soviet delegation to the International Congress of the History of Science held in London in 1931."[26] To counter Bernalism, the SFS set out to elaborate a liberal alternative to the Soviet "planning of science."[27] In the late 1940s, these arguments gained momentum, resonating with the anti-Soviet rhetoric of the early Cold War. In 1946, Baker called on the four hundred or so members of the SFS to renew their attack on Bernalism.[28] The time was ripe, he argued, since "the conflict between the opposite ideals of individualistic freedom and of the highly organized state with its tendency to totalitarian compulsion reached a new degree of intensity."[29]

Bernal's proposal for a centralized British information service became a handy target for the SFS's renewed attack on Bernalism. In May 1948, Baker wrote to *Nature* about the proposal, warning: "The scheme for centralized printing . . . seems to threaten scientific freedom very directly."[30] Major national newspapers—*The Times*, the *National Review*, and the *Observer*—picked up the story, attacking Bernal's ideas as a threat to scientific freedom and liberal democratic ideals.[31] Just as the Royal Society conference opened in late June 1948, the Berlin Airlift began. At this point, Bernal's paper was removed from the conference program, despite his objections. When news of Lysenko's takeover of biology in the Soviet Union broke out later that summer, Baker and Polanyi seized on the fate of genetics in the Soviet Union as exemplifying the threats that Bernalism presented for science. Bernal's public endorsement of Lysenko in 1949 gave these arguments additional weight, ending any hope that his remaining supporters had for a centralized system of scientific information in Britain.[32]

While Bernal felt increasingly isolated in Britain and the United States, he was warmly welcomed in the Soviet Union, where he became quite popular after World War II. He was prominently involved in the Communist-led peace movement and participated in a host of pro-Soviet international organizations.[33] As a participant in the infamous World Congress of Intellectuals in Wrocław in August 1948 (see chapter 4 above), he engaged in a polemical campaign against what he considered a deliberate misrepresentation of the congress in the British press.[34] He took part in the founding of an organization that stemmed from the congress, the World Peace Council, and became the council's vice president. He also established personal contacts with high-level Soviet politicians, including Khrushchev, and was celebrated in the Soviet Union as a "true friend of the Russian people."[35]

Bernal traveled to the Soviet Union almost every year after his first postwar visit in 1949. Each time, he met with the president of the Soviet Academy of Sciences, Sergei Vavilov, and, after Vavilov's death, with his successor, the chemist Aleksandr Nesmeyanov.[36] Both Vavilov and Nesmeyanov were keenly interested in issues of information management, which had been on the agenda of Soviet science planners since the 1930s.[37] Nesmeyanov, in particular, followed British debates on establishing a nationwide center of scientific information.[38] Bernal most likely gave Nesmeyanov the text of his planned presentation at the Royal Society conference since the paper was later widely cited in the Soviet Union and within the Soviet bloc.[39] His plea for a centralized information system resonated with Soviet scientists' own campaign to establish an all-Union center of scientific information.

While the onset of the Cold War put an end to the campaign for a centralized information service in Britain, in the Soviet Union it enabled it. In the aftermath of World War II, Stalin's foreign policy increasingly shifted from a cautious alliance with the West to closing off the Soviet Union to outside influences. By the early 1950s, most channels of scientific exchange between the United States and the Soviet Union were blocked. In the absence of direct contact, scientific publications offered a reliable way to find out about scientific developments abroad, including those pursued in secret.[40] The Institute of Scientific Information, which was opened in Moscow in June 1952, was founded mainly to make *foreign* scientific literature accessible to Soviet specialists.[41] Bernal visited the institute's headquarters—a "large, five-story building on the northern outskirts of Moscow"—in December of that year. Here was "the new nerve centre of Soviet scientific information," he wrote in his travel diary.[42]

By the mid-1950s, the Institute of Scientific Information in Moscow, renamed the All-Union Institute of Scientific and Technical Information (Vsesoiuznyi institut nauchnoi i tekhnicheskoi informatsii, or VINITI) in 1955, quickly became a common reference among Western science administrators, librarians, and documentation specialists. In 1956, when the "documentation consultant" Eugene Garfield offered his "assistance" to the national committee for the planned International Geophysical Year (IGY) "in handling the scientific information and records which will accrue from the [IGY] programs," he referred to Soviet "documentation facilities available to the Russian scientist" as a model "we shall also be hard pressed to duplicate . . . in a few years." "I do not look to such centralized, governmental control as an answer," he continued. "We must, however, fully appreciate the dangers of being left behind." As he saw it: "The time has come for the creation of . . . a scientific body whose scope is the totality of information pertinent to science and research."[43]

Like many commentators in the United States at the time, Garfield—by no means a radical socialist—was fascinated by the possibilities that a socialist state could open for information management.[44] When Garfield met Bernal in the mid-1950s at a conference, he was inspired by Bernal's ideas about information management under socialism. The same year he approached the IGY planners, Garfield advertised Bernal's proposals for a centralized clearinghouse of information in the pages of the *Chemical Bulletin*. "Bernal proposed some time ago," he wrote, "that a centralized reprint clearing house be established. Each scientist would then regularly receive papers in designated areas of interest. The proposal is excellent."[45] A few years later, he tried, unsuccessfully, to pitch Bernal's idea for a newspaper service for science to the National Science Foundation.[46]

It was Bernal who sparked Garfield's interest in the history of science as well. Around the time that Garfield began seeing Bernal at the conferences on documentation and information management they both attended, Bernal finished his *Science in History*, which examined the interplay of science and society from prehistoric times to the present. It originated in a series of lectures Bernal gave at Oxford in 1948, and, according to the recollections of his close friend Aaron Klug, until finishing the manuscript, Bernal regarded this project "as equal in importance to any of the experimental research going on [in his laboratory] at Birkbeck at the time."[47] Given how consumed Bernal was by the preparation of the book, it is not surprising that a young documentation specialist whom he took under his wing turned to the history of science as well. Years later, Garfield would dedicate his life-work "to the memory of the late John Desmond Bernal, whose insight into the societal origins and impact of science inspired an interest that became a career."[48] Yet Garfield's approach to the history of science was quite different from Bernal's. In his private campaign to modernize the management of scientific information, he used historical scholarship on science as a repository of useful data and a testing ground for his company's data products.

Envisioning History as Data Science

If the onset of the Cold War launched a state-supported information campaign in the Soviet Union, it was Sputnik that prompted a similar campaign in the United States. The launch of the earth's first artificial satellite on October 4, 1957, was widely seen as a crowning achievement of Soviet socialism. Within a few months of Sputnik, Garfield changed the name of his information consulting firm in Philadelphia, incorporated in 1956 as Documation (shorthand for Documentation Automation), to the Institute for Scientific Information (ISI), which for many sounded like a nod to the Soviet information center.[49]

As a major military-technological project, Sputnik fueled anxieties that the United States might be lagging behind the Soviet Union in the crucial spheres of scientific research and military strength. Facing an unprecedented public outcry over the "missile gap" and the apparent overall scientific and technological inferiority of the United States, President Eisenhower made science policy the center of his response, launching a host of reforms in the organization and funding of science and education. One of his first actions was the appointment of a high-profile scientific advisory body, the PSAC, to help design a new government science policy.[50] Within the first year of its existence, the PSAC formed a special committee chaired by the physicist Alvin Weinberg to investigate the "problem of scientific information."

For Weinberg, the committee work was one of the contexts of his reflections about Big Science (see chapter 5 above). The report that his committee submitted to the PSAC in 1963 stressed just how much the amount of scientific information available had exploded since the end of World War II. The management of information, however, had not caught up with the pace of its production. The report described the current state of affairs in the United States as a national "information crisis." If a solution to information overload was not found soon, that information crisis would threaten the very identity of science.[51] The report offered a few recommendations, including the wider use of the "mechanical translation of foreign texts" and the "mechanical analysis of texts for indexing and retrieval." It also singled out a new tool, the Science Citation Index (SCI), that Garfield had been developing at the ISI since the late 1950s. According to the report, the Weinberg Committee had been "particularly impressed" with the potential of computer-assisted citation analysis when it came to retrieving and evaluating scientific information.[52]

In his first publication on the subject, in *Science* in 1955, Garfield described the citation method as "a new approach to . . . the literature of science."[53] At its core was genealogical reconstruction. Its basic idea was to produce a pedigree by tracing a current scientific idea or method or some other scientific intervention back to its oldest ancestor as it was expressed in the literature through citations. Garfield stressed that the success of the method depended on whether a "large volume of data" could be handled by "mechanical devices of high speed and versatility," which would not only facilitate this work but "probably determine success or failure." The result would be a citation index containing the information ancestry—"the bibliographic descendants of antecedent papers"—a retrieval tool that would allow researchers to identify relevant publications on specific subjects quickly and keep abreast of the literature across disciplines. It could also function as an evaluation device to gauge the impact of a particular author, article, or journal over time.[54]

The *Science* article helped promote Garfield's proposal in the academy. One of the scientists to take early notice of it was the geneticist Joshua Lederberg, who participated in the development of the SCI and put it on the agenda of the Weinberg Committee, which he joined at Weinberg's invitation as the committee's only biologist.[55]

Bernal, who had joined the SCI advisory board, lent his full support to the project. In a laudatory review, he particularly emphasized that Garfield's information-genealogy method was the kind of symbiosis of science history and science policy that he himself had been trying to forge in his *Science in History*. "The value of the Science Citation Index was immediately apparent to me because I had tried to do the same thing in reverse order in writing about various aspects of the history of science," he mused. "Its essential value is . . . that it is a new dimension in indices which should enable the polydimensional graph on the progress of science to be mapped out for the first time. Such a graph is a necessary stage in drawing up or planning any strategy for scientific research as a whole."[56]

In his correspondence with Lederberg, Garfield mused that the new breed of information specialist would be a computer programmer, well versed in the techniques of machine translation, mechanical indexing, and citation analysis. Along with these skills, a "solid training in science history" would be a must.[57] Garfield's coauthored *The Use of Citation Data in Writing the History of Science* (1964) gives a glimpse of how the history of science became entangled and reconciled with informatics at midcentury.[58]

Garfield's underlying assumption was simple: he reckoned that footnotes and bibliographies in existing historical accounts of important scientific breakthroughs could be used as ready-made databases that could be mined to construct chronological maps and the "topological network diagrams."[59] In his *Science* article, he noted one of the applications of the envisioned citation index: "[The SCI] would be particularly useful in historical research, when one is trying to evaluate the significance of a particular work and its impact on the literature and the thinking of the period."[60] In fact, he saw the relation between historical scholarship and the SCI as a two-way street. For historians, assembling bibliographies on any given subject is second nature, a necessary component of history writing. Featured in every historical publication, bibliographies are scrutinized by peers to determine authors' competence and the profundity of their work.[61] For information scientists, on the other hand, historical scholarship was a trove of data that could be examined using new citation analytics tools. Scholars, too, examine bibliographies or footnotes, looking up specific references of interest, but they do not analyze them collectively as meaningful entities. Garfield's team set out to do precisely this. They used historical scholarship as a repository of data—citation data—to test the new tool.[62]

To demonstrate the possibilities of the citation method, Garfield approached the popularizer of science and science fiction writer Isaac Asimov with the idea of using his 1962 *The Genetic Code* for an experiment in scientific historiography.[63] He proposed to compare two methods of writing history, or as the team put it in the book: "(1) conventional or traditional subjective analysis, (2) objective citational or bibliographical analysis."[64] *The Genetic Code* was to represent the former. The fact that the work was not intended as a formal contribution to the history of science did not deter him. What mattered was that, in his account, Asimov set out to trace, step-by-step, the history of genetics, starting with Gregor Mendel's classic pea experiments, rediscovered in

1900, to a major breakthrough in contemporary biology—the cracking of the genetic code by Marshall Nirenberg and his colleagues at the National Institutes of Health in 1961 when, after years of intense search for the "alphabet of life," molecular biology received its Rosetta Stone.[65] As Asimov explained in the introduction of *The Genetic Code*, his aim was "to explain the background of the breakthrough; the full meaning of the breakthrough and its immediate consequences; and, finally, [to provide] a forecast of what the breakthrough may bring about in the future—what the world of 2004 may be like."[66]

Having received Asimov's consent, Garfield and his team first mined *The Genetic Code*'s bibliography to identify the most important publications leading to the discovery (they called them the *nodal articles*). Next, they constructed a network diagram showing the "historical dependencies" between the nodal articles. Then, they examined the bibliography of each nodal article to determine who cited whom. If a particular node could not be linked to any earlier node by direct citation, they searched other articles for indirect connections. Then, they used the SCI, which they had just finished compiling, to determine "the total count of first author citations to every work for each nodal author." Using the citation data, they constructed another network diagram. Finally, they mapped and compared "the interdependencies of linkages among 40 major events (nodes) included in both network diagrams."[67]

The result was the two networks representing the two methods—one visualizing the historical dependencies as presented in *The Genetic Code*, the other being Garfield's historiogram presenting citation method in action. The two networks turned out to be fairly similar (see fig. 5). The citation method yielded a remarkably predictable who's who of DNA research, reproducing the conventions that excluded women, technicians, and younger researchers from the DNA story. For instance, Rosalind Franklin, whose work was crucial for James Watson and Francis Crick's discovery of the DNA structure, did not appear in either account.[68] The contributions of marginalized scientists, indeed, more often than not do not leave a measurable impact in the form of others citing their work. The citation diagrams reproduced their exclusion with graphic precision.

Garfield was aware of the dangers of overinterpreting his results, and he was savvy enough to market his method differently to different groups. To historians, he pitched the citation method as both an information retrieval tool and a visualization method allowing researchers to grasp a discipline's trajectories quickly and "identify key events, their chronology, their interrelationship, and their relative importance." "Unquestionably," he and his team argued, citation networks revealed more

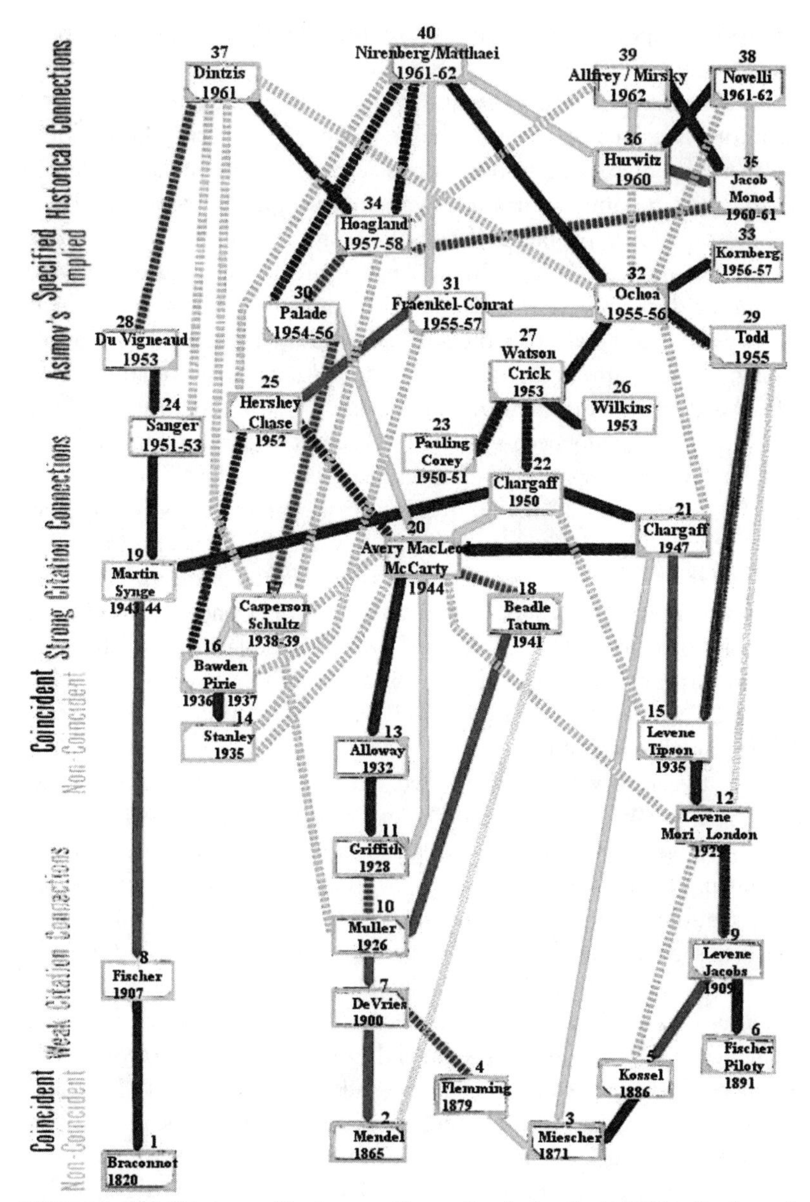

FIGURE 5 Garfield's historiogram of the discovery of the genetic code: from Mendel (1865) to Nirenberg and Matthaei (1961–62). Eugene Garfield, Irving H. Sher, and Richard J. Torpie, *The Use of Citation Data in Writing the History of Science* (Philadelphia: Institute for Scientific Information, 1964), app 8.

"historical dependencies" than historians could possibly narrate in their texts.[69]

In *The Use of Citation Data*, Garfield also pitched computers as a tool for historians of science. The book opened with a bold question: "Can a computer write the history of science?" As the authors explained in the introduction, their study was an "attempt to understand and define some basic problems if computers are ever to aid the historian of science—no less supplant him." They argued: "The citation network technique does provide the scholar with a new *modus operandi* which, we believe, could and probably will significantly affect future historiography." History, they reasoned, was a literary enterprise, and the history of science was no different in this respect: "Writing the history of science has traditionally been a purely intellectual or cerebral pursuit of the scholar." Computer-assisted citation analysis using the SCI represented "useful heuristic tools" that allowed the historian to "employ new techniques for sifting and evaluating data," revealing historical connections through the analysis of cross-references and citation patterns.[70]

In the 1960s and 1970s, many historians of science added citation analysis to their analytic tool kits.[71] Derek de Solla Price, a physicist-turned-historian of science at Yale, was instantly hooked by Garfield's method. Garfield had reached out to Price in 1962. Along with his letter, he sent a collection of reprints and "sample indexes."[72] An enthusiast of quantitative history, Price was thrilled. "I am strangely excited by the material you sent me on the Citation Index Project," he wrote back.[73] In *Science since Babylon*, a collection of essays published in 1961, he laid out an argument about the interrelated exponential growth in the number of scientists, number of journals, national investment, and scientific literature.[74] Garfield's letter arrived when he was working on his next book, *Little Science, Big Science*, in which, using available statistics, he set out to develop his intuitions about the growth of science.[75] "What I want now is some numerical analysis, however rough, of the rate of exponential growth," he wrote to Garfield.[76] An intense correspondence followed, with Price asking for more data and giving generous advice about how best to set up the SCI to make it work as a multipurpose "calculus of science."[77]

Another historian of science found Garfield's method congenial: Thomas Kuhn. The notions of "paradigms," "normal science," "revolution," "anomalies," and "crisis," which Kuhn introduced in his famed 1962 *The Structure of Scientific Revolutions*, lent themselves to visual schematizations of the kind made possible by Garfield's method.[78] In the postscript to the 1969 edition of *Structure*, Kuhn discussed Garfield's method, responding to critics who accused him of "trying to make science rest on unanalyzable individual intuitions rather than on logic and law." He cited *The Use of Citation Data in Writing the History of Science* as an example of how the structure of scientific communities could be quantitatively analyzed through visualizing and measuring the "linkages among citations." He noted that he was "currently experimenting with a computer program designed to investigate [citations'] properties at an elementary level."[79]

Historians and Computers

Garfield's efforts to pitch the citation method to historians of science coincided with the resurgence of quantitative history, promoted by new computing technologies. Historians who were already engaged in quantitative history were quick to explore the possibilities of using computers to delegate the tedious work of organizing, sorting, retrieving, and analyzing large amounts of data. Scholars associated with the *Annales* school were in the vanguard of computer-aided historical research, and, not coincidentally, one of the first historical theses that used an electronic computer for quantitative work was produced by a French historian. In 1969, Marcel Couturier, a recent graduate of the Sorbonne, used its IBM S/360 computer in his study of the historical demography of Châteaudun, a small city southwest of Paris.[80]

By 1964, when IBM introduced the S/360 family of computers—designed to cover a wide range of applications—it had already established a solid presence in France, opening up sales offices in many French cities. While DeGaulle's government was concerned that the growing American presence could put French computing technologies at a disadvantage, in 1965 it authorized a major contract between the French Department of Defense and IBM for the establishment of a research facility, Special and Military Systems, at Essonnes, to design systems for the French army, navy, and air force. Other ministries and French universities followed their lead and acquired IBM mainframe computers at this time.[81]

Couturier's use of the computer did not alter his approach to data gathering and analysis, which was informed by his training in historical demographic studies. This field had become prominent in France in no small part because of the *Annales* school, which, as we saw, promoted the practice of gathering demographic data region by region since the 1930s (see chapter 5 above). In the 1960s, the methods of large-scale demographic history were standardized. A widely used field manual advised research teams to use standard color-coded forms for a prearranged sampling study. The forms were distributed to provincial universities, where coordinators passed them along to students and research assistants assigned the task of gathering data in remote villages. After collecting a large sample of homogeneous data from the completed forms, the historians collated, sorted, and aggregated the data, often using punch cards and mechanical card sorters.[82] In a procedure not altogether different, Couturier punched about sixteen thousand cards, each representing a single data point—the registration of a marriage, a birth, an act of apprenticeship, etc.—and simple programs were written to analyze the data over different time periods, both for the city as a whole and for particular parts of it.[83]

Historians who were trained to do this kind of work readily embraced computers as tools to crunch large bodies of data more efficiently and quickly. In 1968, a leading demographic historian in France, the Annaliste Emmanuel Le Roy Ladurie, predicted that, by the 1980s, "the historian will be a programmer or he will be no more."[84] Even Braudel, who like other Annalistes used economic statistical methods in his work, expressed a cautious belief in the "utility" of the continually improved "calculating machines." Computer-assisted "social mathematics," he thought, might push *longue durée* studies "ever further back in time."[85]

From the perspective of a contemporary historian in the United States, Braudel treated computational history as a "mistress who fascinates but is not taken too seriously."[86] Indeed, the historians who launched the New Economic History in the United States were taking mathematical approaches to history quite seriously. When, in the mid-1960s, a group of French historians began to follow what they referred to as the *American approach* to economic history, they vigorously criticized their compatriots for having little awareness of how to conduct a rigorous statistical analysis, build mathematical models, or test hypotheses explicitly and rigorously.[87]

The New Economic History movement was boosted by the increased availability of general-purpose computers, which were marketed to a wider range of users.[88] In 1965, a year after delivering the first in its S/360 family of inexpensive general-purpose computers, IBM sponsored a new journal, *Computers and the Humanities* (*CHum*), an initiative of a professor of English at the City University of New York, Joseph Raben.[89] *CHum* targeted academics across the humanities at precisely the time that universitywide computing facilities were springing up throughout the United States.[90] It regularly advertised IBM products, publishing reviews of available software, beginning with a list of "computer programs designed to solve humanistic problems."[91] Addressing his fellow historians in the first issue, the historian of medicine Vern L. Bullough stated: "Computer analysis is not so difficult or foreign as it might at first seem since it really is an extension of ideas and concepts already inherent in the historical methodology."[92] Some historians found it worth the effort.[93]

Many traditional historical problems and questions do not, however, lend themselves to computation or even quantification more generally. Even those historians and social scientists who became early computer converts wanted more flexible tools that would allow them to apply the convenience of computer-assisted research to their work with texts, lists, maps, catalogs, and other nondigital renderings of data.[94] Many more were simply not interested in adapting their approaches to computer analysis.

The new scientific turn in history, now promoted by the arrival of general-purpose computers, prompted a quick and powerful backlash from leading historians disconcerted by the trend. In 1960, the Oxford don and iconic anti-Communist intellectual Isaiah Berlin offered an extended critique of scientific history in an essay that inaugurated a new journal, *History and Theory*. Berlin used an image of a computer to deride a "crav[ing] for a natural science of history" and asked whether "electronic brains" would make a new brand of scientific history any different in kind from prior attempts that hoped to "do away with the chaotic accumulation of facts, conjectures and rules of thumb" characteristic of "the uncertain, old-fashioned, hand-operated tools with which historians had fumbled their way in the unregenerate past?" His response was a resounding no. In his account, the nineteenth-century historian's dream of "revolutionis[ing] the old empirical, hand-woven history by mechanizing it, . . . [just] as Newtonian physics transformed older cosmologies," remained just that—a futile and misplaced dream: "The great machine which was to rescue [historians] from the tedious labours of adding fact to fact and of attempting to construct a coherent account out of their hand-picked material, remained no more than a plan in the head of a cracked inventor . . . a bogus prospectus."[95] When, five years later, Berlin was approached by the UNESCO History of Mankind team as a potential reviewer, he swiftly declined.[96] Had he agreed to provide commentary, he likely would have been as critical as the Cambridge historian John H. Plumb, who, as we have seen, ridiculed the project's work as "an encyclopedia gone berserk, or re-sorted by a deficient computer" (see chapter 5 above).

In the early 1960s, many historians lambasted quantification as dehumanizing and threatening the very essence of historical scholarship.

In his 1962 presidential address delivered at the annual meeting of the American Historical Association, Carl Bridenbaugh told his fellow historians: "The finest historians will not be those who succumb to the dehumanizing methods of social sciences. . . . Nor will the historian worship at the shrine of that Bitch goddess, QUANTIFICATION."[97] By the late 1960s, when the punch card had become a symbol of the uniformity and depersonalization of the post–World War II knowledge culture for many people within and beyond countercultural movement, even those forms of quantification that grew from within the discipline itself had largely been rejected by the historical profession.[98]

By this time, Garfield's efforts to promote the SCI likewise had become the subject of controversy and complaint. Some critics mocked the citation analysis method as premised on the assumption that "objectivity [can be] gained by not reading the literature."[99] For Garfield, a bigger problem was that his commercial corporation, the ISI, was struggling to make ends meet. Compiling the index proved to be more expensive than anticipated, and, because of the SCI, which did not sell well, the ISI had not yet generated a profit and almost went under.[100] The SCI was a promising product, but the market for it did not yet exist; it had to be created. Under these circumstances, Garfield turned to the socialist market for a solution.

The Socialist Market for a Capitalist Data Product

In the aftermath of Stalin's death, East-West scientific exchanges, halted since the onset of the Cold War, resumed. In the Soviet Union, restrictions on travel abroad were eased. By mid-1950s, Soviet scientists in various fields—in particular, informatics and computer research—became regular participants in international conferences in the United States.[101] In the United States, the post-Sputnik climate prompted the Eisenhower administration to establish a legal framework for broad areas of educational, cultural, and technical exchanges with the Soviet Union.[102] Starting in the early 1960s, delegations of librarians began crisscrossing the Atlantic, joining delegations of other specialists and administrators.[103] In 1964, a new agreement expanded the scope of exchange to include "information specialists" working in "special libraries" and other centers dealing with information management. Under the terms of the agreement, one delegation from each side visited the other's facilities to study techniques for compiling bibliographies, indexes, and other information resources.[104]

In 1965, a delegation from the Soviet Union visited the ISI in Philadelphia as one stop on its tour of US information centers. The delegation included A. M. Mikhailov, the director of VINITI. By the time of Mikhailov's first official visit to Garfield's company, the two must have known each other fairly well. Mikhailov had been on the editorial board of the SCI since Garfield's first visit to VINITI in 1961, when he came to the Soviet Union on a private tour.[105] This official visit by a high-profile Soviet delegation, however, helped publicize the ISI as an important US information center, comparable to the Library of Congress and other institutions visited by the Soviet delegation.

The visit became an occasion for science journalists in the United States to reassess the popular perception of the Soviet approach to information management. Thus, Gary Brooten, a reporter for the *Philadelphia Inquirer*, interviewed members of the delegation, including Nikolai Arutyunov, the chief of the Administration of Scientific and Technical Information and Dissemination of the Soviet State Committee on Science and Technology (Gosudarstvennyi komitet po nauke i tekhnike, or GKNT), a newly established science policy-making organ. He reported that Arutyunov agreed with his assessment of VINITI as "the largest operation of its type anywhere." Quoting Arutyunov, the account detailed how VINITI covered some 2.5 million articles published in eighteen thousand journals in "virtually the entire field of science" annually. However, Brooten noted: "It is not true, as some reports have suggested, he [Arutyunov] said, that the Soviets have tried to solve the problem [of information management] by putting everything they know into a single computer center and hooking it up electronically with libraries throughout the country." He quoted Arutyunov's response regarding the "information crisis" in the United States, a problem that the Soviets had supposedly escaped: "Russia has about as much trouble as the United States keeping its scientists informed of developments in their fields, a Soviet expert on information sciences said. . . . 'The differences between our system, which is supposed to be centralized, and yours, which is supposed to be decentralized, are mostly matters of terminology,' Arutyunov said."[106]

By the time of Mikhailov's visit, Garfield's own perception of the Soviet model of information management had shifted as well. His earlier admiration of VINITI as the embodiment of centralized socialist planning in the sphere of information gave way to a more pragmatic assessment of Soviet information centers as markets for his products. Using his contacts in VINITI and other information centers and the research libraries in the Soviet Union and Eastern Europe, Garfield began negotiating sales of the SCI and other ISI data products, such as the scientific literature digest *Current Contents*. Following the official Soviet visit to Philadelphia, Garfield pitched his products to Arutyunov. "It seems to me," he wrote to Arutyunov, "that it is far cheaper in the long run for Soviet scientists and librarians to utilize our existing services than to attempt a duplication of them."[107]

A skilled salesman, Garfield packaged the subscriptions together with the sales of IBM computers. This commercial strategy took advantage of the shortage of general-purpose computers in the socialist bloc. By the late 1940s, the Soviet Union had a small but independent computer industry producing machines used to control conventional and, increasingly, nuclear weapons and ballistic missiles.[108] But, in the early 1960s, Soviet strategy had shifted from the development of its own, largely domestically produced line of computers to the direct acquisition of IBM machines and other forms of Western technology for reverse engineering and copying. The Soviets' access to Western computer technology was, however, limited because NATO banned the export of so-called dual-use technologies—civilian products with military applications—to Eastern bloc countries.[109] Consequently, Soviet production of computers was falling behind US and Western European production, and the Soviet demand for IBM computers was strong.

The diplomatic normalization of East-West relations associated with détente lifted some of the restrictions previously imposed on the sale of dual-use technologies by members of the NATO bloc to the Soviet Union and members of the Soviet bloc. In 1967, NATO adopted a new strategy (the so-called Harmel Doctrine) that approved a cautious reopening of US-Soviet trade. Diplomatic talks over the next few years culminated in a 1972 trade agreement between the United States and the Soviet Union establishing a legal framework for US-Soviet trade and lifting restrictions on earlier models of IBM computers, the exemplary dual-use technology. Even before the trade agreement was officially signed, IBM started talks negotiating the sale of its 360/50 and 360/40 computers to several Soviet ministries.[110] Garfield acted as one of the mediators between IBM and Moscow.

The strategy of making the sale of ISI products part of a larger deal including IBM machines proved to be a gold mine for Garfield. In the summer of 1971, Garfield brokered the first sale of an IBM 360/40 computer to VINITI. As part of the $2 million deal, he also negotiated "a multiyear arrangement for 5 to 10 years duration . . . of ISI publications, services, magnetic tapes and software." The ISI fees—$250,000 per year—were folded into the cost of the package.[111] Other commercial agreements followed, each featuring subscriptions to the SCI as part of the "information package."[112] Cooperation between the ISI and VINITI expanded throughout the 1970s and 1980s, when VINITI and a host of research institutes in the Soviet Union purchased the SCI and other information products. While VINITI did maintain a subscription to the SCI, it was paid for by means of what Garfield characterized as "mutually beneficial barter agreements." As payment for ISI data products, he was willing to accept "Russian manpower" for "some specific work which Soviet scientists could do for the ISI."[113]

For its part, VINITI appeared to be an eager consumer. Indeed, from its very beginning, its leadership promoted the use of computers. Its first director, Dmitrii Panov, was the deputy director of the Institute of Mechanics and Computational Technology before he was appointed to head the new institute of scientific information. One of Panov's first actions as director was to produce a report on the state of and tendencies in the development of the computer industry in the United States that was based on an extensive review of published sources. The report, submitted to the Party's Central Committee in March 1955, stressed American advances in the field and the Soviet "lag," in terms of both the quantity and the quality of general-purpose computers. The report also pointed out that security experts could glean precious clues about strategically important (and often classified) developments in the West through surveying and analyzing scientific and technical publications available through open access.[114] During the 1960s, the emphasis on computers and the automation of information processing became a recurrent theme in the public statements of VINITI planners. After a British delegation visited VINITI's facilities in 1963, its laudatory report on the visit cited one of VINITI's architects, Nesmeyanov, extolling automation and machine methods of information management as "the future path of scientific information."[115]

When Garfield visited VINITI for the first time in 1961, he was eager to see how the Soviets used automation. On his arrival, however, he did not find much automation occurring.[116] Instead, he found a very large number of full-time employees who were processing information by hand, sometimes with the help of electromechanical devices.[117] The envy of science planners and information specialists in the West who marveled over the possibilities in a socialist system to organize information efficiently, VINITI was, indeed, a socialist institution, in one specific sense: it provided job security to anyone who was professionally qualified. Owing to its demand for personnel with some scientific training and knowledge of foreign languages, it was a safe haven for the many Jewish scientists who lost their jobs during the "anticosmopolitan" (i.e., anti-Semitic) campaigns of the late 1940s and early 1950s. It also employed many scientists who held politically dissenting views as well as scientists who had been sent to the Gulag after the Great Purges of the 1930s and, after spending years in the labor camps, were released after Stalin's death. Not employable in academic institutes, many of them landed at VINITI.

Vasilii Nalimov, who played a crucial role in publicizing the citation method in the Soviet Union, was in the latter category. Born in 1910 into a professor's family, he began studying physics and mathematics at Moscow University soon after the Revolution. His interests soon moved beyond physics, and he dropped out of university. As a student, he had begun to attend an informal literary-philosophical seminar led by Alexei Solonovich, a professor of mathematics and the spiritual leader of the Russian "mystical anarchists."[118] In 1930, Solonovich was arrested; by 1936, Nalimov and the remaining members of the seminar were arrested as well. Nalimov received a sentence of ten years of forced labor for "propagating mysticism" and was sent to the Kolyma camp. He cut wood in Kolyma for eight years before being moved to a *sharashka*, a secret research and development lab in the Gulag labor camp system.[119] From there, he was reassigned to a metallurgic plant in Dzheskazgan in Kazakhstan mostly staffed by Gulag prisoners. There, as the head of quality control, he resumed his mathematical work and began applying probabilistic methods and statistical techniques to the monitoring and controlling of product quality. He published his work in the local journal *Kolyma*.[120] After the war, he began publishing in the more widely distributed *Industrial Laboratory* (*Zavodskaia laboratoriia*). On his release shortly after Stalin's death, he returned to Moscow. With his three foreign languages and stock of remarkably competent publications in *Industrial Laboratory*, he was able to land a job at VINITI as an abstractor.

Nalimov hit on the SCI by accident. As he recounted in his autobiography, one day in 1958 or 1959 he received an article in Italian by Derek de Solla Price for abstracting. It was one of many hundreds that he abstracted during his time at VINITI, but he found this one, on the exponential growth of science, interesting. He thought that he could take Price's argument one step further and find a formula for his curves. His project soon changed, however, as he was struck by the notion that science could be regarded as a cybernetic information process in which publications functioned as units of information. With two other researchers at VINITI, he coauthored an article discussing the idea, publishing it in one of the leading Soviet physics journals, *Uspekhi fizicheskikh nauk*.[121]

According to Garfield, he met Nalimov during one of his first visits to the Soviet Union in the early 1960s when he was promoting ISI products at the Moscow book fair. Nalimov, who stumbled on Garfield's booth, introduced himself. The two immediately bonded.[122] By then, Nalimov was no longer with VINITI. After a brief stint at a chemical research institute in Moscow, where he worked on developing a mathematical theory of experiment (the work he started while in the Gulag), he had been invited by the world-renowned mathematician Andrei Kolmogorov to join the interdepartmental Laboratory of Statistical Methods at Moscow University as his deputy.[123] When he met Garfield and received a promotional copy of the SCI, he reasoned that the citation method could be useful for his work.[124]

Nalimov started an informal seminar on citation analysis in order to explore its applications. The group started by reverse-engineering the published SCI. The participants, most of them members of the Chemistry Department of Moscow University, divided the work among themselves, each one taking responsibility for a particular field—analytic chemistry, spectrum analysis, chemical physics, and so forth. They examined the growth and distribution curves of publications in their areas by journal and by country, complementing the data they found in the SCI with their own analysis of publications in Russian. The work was done by hand: in the evening hours, they rifled their way through journals in the libraries, searching for references. They counted publications and citations using a mechanical blood cell counter.[125]

The seminar resulted in a coauthored article summarizing the group's work and featuring detailed descriptions of how to use the SCI to retrieve information on English-language scientific publications, evaluate particular ideas, and analyze larger patterns. The article also highlighted how incorporating publications in Russian changed the overall picture of "world information circuits." Their study, the authors stated, "established a modest influence of Soviet science on world information flows." They suggested that this relatively limited influence could be attributed to one of two possible explanations: Soviet science had no impact either because the information pipelines were clogged (i.e., Soviet journals were too slow in publishing ongoing scientific research) or because information flows in English and Russian existed as two separate information circuits, with the Russian-language information flow not linked to the world information circuit.[126] It was left to the reader to decide which explanation was more likely.

The article led to the institutionalization of the seminar under the aegis of the Institute for the History of Science and Technology in Moscow (Institut istorii estestvoznaniia i tekhniki, IIET) as a continuing colloquium on the quantitative methods of science studies (*naukovedenie*).[127] It became popular and introduced a wide range of academics to the citation method.[128] Nalimov's *Naukometriia: Izuchenie razvitiia nauki kak informatsionnogo processa* became a primer on the citation method. It also publicized the SCI, the main tool of scientometrics, in the Soviet Union and the Soviet bloc countries.[129]

For the most part, the early champions of the citation index method in the Soviet Union had very limited access to computers, with one important exception. In the late 1960s and 1970s, it was taken up by specialists in computer-based forecasting of long-term economic and social trends.[130] Gennadii Dobrov, a forecaster who was the deputy director of the Kiev Institute of Cybernetics, adopted the citation method to analyze trends in *unpublished* literature. Working in close contact with the GKNT, he used the reports deposited at the Center of Scientific and Technical Information (Tsentr nauchno-tekhnicheskoi informatsii, TsNTI) as his data.[131] Unlike VINITI, which functioned as a clearinghouse for published information, TsNTI was a depository housing limited-access "gray" literature—research reports that scientists from all over the Soviet Union submitted as preprints. The reports could be made available to other scientists on request. Dobrov tracked the requests the same way Garfield tracked citations. Like Garfield, he advocated using computers in the history of science, mining scientific bibliographies to build forecasting scenarios. In the late 1960s, he established the Center of Computer Methods of Processing Information in the History of Science for this purpose.[132] Like Nalimov, he publicized Garfield's data products in the Soviet Union. Dobrov met Garfield in 1966 when he invited the Philadelphia entrepreneur to a symposium on scientific forecasting that he organized in Kiev. The symposium became an annual event, with Garfield usually in attendance.[133]

When, in the early 1970s, Garfield was negotiating his $2 million deal with VINITI for an IBM computer and the accompanying "information package," the socialist market for the SCI was second only to that in the United States. In 1980, the ISI moved into a new, multimillion-dollar building in the center of Philadelphia. By then, Garfield himself had become a millionaire. Ironically, the SCI, built with computers and for computer users, became a commercially viable product owing in no small part to the labor of its dedicated Soviet users, who had very limited, if any, access to computers and counted citations by hand or with the aid of simple mechanical devices. Commercial interests played an important role in the development of the SCI. They are clearly revealed in the financial transactions and the deals that Garfield brokered with his Soviet partners. But equally important were the ideologically laden imaginaries of the Other. The development and promotion of this new data-analytic tool, the SCI, was driven by both socialist markets and socialist visions. Garfield used both to great effect.

By the time the SCI became a commercially viable product, quantitative methods in history were under fire in the United States. In the wake of the culture wars of the late 1960s, computers emerged at the polemical core of the countercultural movement.[134] Student protesters targeted the use and misuse of computers in managing the growth of large institutions, including their own universities and the military-industrial complex.[135] Along with university think tanks operating on defense contracts, protesting students took computer centers hostage as symbols of what they saw as science's complicity in subverting public policy, eroding constitutional control, and threatening democratic institutions.[136] The fading appeal of quantitative history was part of a larger questioning of Cold War rationality—the commitment to rationalistic, technical, and instrumental approaches and methodologies that flourished in the political climate of the 1950s and the early 1960s.[137]

In the Soviet Union, computers became entangled in different culture wars. At the onset of Khrushchev's thaw and the de-Stalinization movement in the 1960s, computers—and science in general—emerged as a remedy to the falsehood of the Stalinist era and the failings of Communist leaders and Marxist theoreticians. In the context of those culture wars, the rationality and objectivity of the scientific method emerged as a way to renew the socialist project. The algorithmic rationality of computers, it was hoped, would salvage socialism from the political arbitrariness of the Party gerontocracy and calcified late Soviet Marxism, protect against the fallacies of the Soviet political leadership, and even open a possibility of building a form of "electronic socialism" by establishing a network of computer centers for centralized economic planning and social management.[138] Even though the network of computers was not realized, a large network of supporters of the program of Soviet electronic socialism among scientists, the media, and the government created a new political imaginary in which the computer stood at the forefront of the post-Stalinist Soviet socialist project.

Not surprisingly, in the climate of the political thaw, attempts to put the computer at the forefront of the humanities and human sciences in the Soviet Union were under way. A stunning early success was the deciphering of Mayan inscriptions, an outcome of a collaboration between the Egyptologist Yuri Knorozov and a team of computer scientists in Novosibirsk that dated to the early 1950s.[139] *Vestnik istorii mirovoi kul'tury*, the journal established as part of the Soviet contribution to the UNESCO History of Mankind project (see chapter 5 above), brought Knorozov's work to international audiences.[140] In 1968, the Academy of Sciences established a special commission to study the application of mathematical methods and electronic computers to historical research. The same year, the Department of History at Moscow University opened a lab on mathematical methods and computers in history (the Laboratoriia po primeneniyu matematicheskikh metodov i EVM v istoricheskikh issledovaniiakh).[141] By the 1970s and 1980s, there were book-length studies of the intersections of history and mathematics.[142]

The collapse of the Soviet Union in the early 1990s turned Soviet-trained mathematicians into post-Soviet entrepreneurs of all sorts.[143] Some of them made forays into rewriting history. In a fitting finale to the story of scientific history, a story replete with ironies throughout the twentieth century, post-Soviet sensibilities make the Russian locale once again particularly receptive to the project of scientific history at that century's end.

Epilogue

This book opened with a moment at the beginning of the twentieth century when concerns about the future of history preoccupied scholars around the world. By the close of the century, many scholars thought that history itself had come to an end. In February 1989, the Sovietologist Francis Fukuyama, who had just left his job at the RAND Corporation to take a position as the deputy director of policy planning at the US Department of State, delivered a talk at the University of Chicago in which he famously claimed that the imminent collapse of the Soviet Union would put an end to history. "What we may be witnessing is not just the end of the Cold War," he wrote in the published version of his talk, "but the end of history as such: that is, the end point of mankind's ideological evolution and the universalization of Western liberal democracy as the final form of human government." He explained that he was not suggesting that the future would be devoid of anticapitalistic movements or that socialistic programs would not reemerge around the world. His point, rather, was that "it [now] matters very little what strange thoughts occur to people in Albania or Burkina Faso, for we are interested in what one could in some sense call the common ideological heritage of mankind." "At the end of history," he explained, "it is not necessary that all societies become successful liberal societies, merely that they end their ideological pretensions of representing different and higher forms of human society." With the end of Soviet socialism, the most serious alternative to liberalism has "cease[d] to be a factor" in the "post-historical world."[1]

Fukuyama gave his talk just two months after Mikhail Gorbachev, speaking at the United Nations on December 7, 1988, announced dramatic changes to Soviet domestic and foreign policy as part of his reform program, perestroika. To many, the unilateral cuts in Soviet military expenditures and a major drawdown of the Soviet military in Eastern Europe and the European part of the Soviet Union signaled the beginning of the end of the Cold War. It must have been an interesting moment for America's Soviet experts, pushing them to think big.

For another Russia expert, the Australian historian David Christian, the impetus to think big took the innovative form of Big History. In early 1989, Christian offered his first Big History course at Macquarie University in Sydney. It reached courageously across the science-humanities divide, pushing the story back to fifteen billion years ago, and bringing it up to the present. In discussing the rationale for such a course in a short publication reporting on the experience, Christian noted that he was driven by the belief that historians needed to reclaim their traditional role as global storytellers. In his telling, the purpose of Big History was to establish a modern creation myth—a secular creation story for the modern world.[2] In 1991, in a programmatic article advocating Big History, he wrote: "History could play as significant a role in modern society as traditional creation myths have played in nonindustrial societies; but it will do so only if it asks questions as large and as profound as those posed in traditional creation myths."[3]

Big History found dedicated followers in Russia. In the early years of the twenty-first century, Christian's brief on behalf of Big History was translated into Russian and published in a journal devoted to post-Marxist approaches to the social and human sciences.[4] Christian's translator, the Armenian scholar Akop Nazaretyan, a Soviet-trained philosopher who had been publishing on cybernetics approaches in social sciences since the 1970s, became one of the promoters of Big History in Russia. In 2011, he established a Big History center, the Euro-Asian Center for Megahistory and Systems Forecasting (Evro-asiatskii Tsentr mega-istorii i sistemnogo prognozirovaniia), in Moscow, serving as its director until his death in 2019.[5] The center popularized Big History in Russia and promoted it globally.[6] In a condolence message published on the center's website, Christian paid tribute to "an old friend and a 'bogatyr' [an old-fashioned Russian warrior] of big history."[7]

It might be a coincidence that Big History was pioneered by a Russianist in the tumultuous year of 1989, but it is hardly coincidental that the approach found a receptive home in postsocialist Russia.[8] The collapse of the Soviet Union traumatized millions of people, disrupting their values, orientations, and cultural identities.[9] It also created a demand for new values and meanings, including the meaning of Russian and Soviet history. Amid the revelations of the violence of the Soviet past as well as the complicity of the Soviet historical guild that failed to record it, professional historians found themselves pushed to the sidelines.[10] As millions of Russians sought new meanings and a new value system in history, alternative histories emerged as a popular forum for new (and sometimes wild) historiographic ideas and concepts.[11]

The most prominent effort in alternative history in post-Soviet Russia has been the New Chronology movement.[12] It was pioneered by a renowned mathematician at Moscow University, Anatolii Fomenko, who began his foray into history in the 1970s and published his first work on alternative history in VINITI's informatics and semiotics series.[13] Using computers and statistical analysis, Fomenko mined data from classical texts, starting with Ptolemy's *Almagest*, an important pre-Copernican astronomical treatise from the second century CE that cataloged the positions of stars and established a model for a geocentric universe. Calculating statistical correlations between the dates of dynastic rules and astronomical events, Fomenko came to the conclusion that the *Almagest* should be dated to a much later period, between the seventh and the thirteenth centuries. He began digging more widely and argued that what historians thought of as ancient history was merely a reflection of events that had occurred much later—that the works attributed to classical Greece and Rome were actually medieval and Renaissance fabrications. As he put it in one of his publications: "Antique events are phantom duplicates, reflections of the medieval originals."[14]

While Fomenko inspired a small cohort of followers among mathematicians, until the 1990s the New Chronology was an idiosyncratic and isolated phenomenon. The void created by the dissolution of the Soviet Union provided a new context for it. In the 1990s, education officials were considering incorporating it in Russian schools' curricula. Since then, hundreds of thousands of copies of books exploring the New Chronology were published and sold in Russia. Fomenko and his followers extended the New Chronology to cover the entire world, folding global history into a new and fantastic temporal landscape.

On the surface, the New Chronology would seem to be the very opposite of Big History: while the former drastically shortens the scale of human history, the latter's goal is to extend it as far as possible. There are, however, some interesting parallels between the two phenomena. Both offer radically new temporal landscapes of history, and both narrate versions of history that can feel *posthistorical*. The New Chronology offers a posthistorical view of Russian history by displacing the Soviet past and presenting medieval Russia as a golden age.[15] For the observers of the Big History movement in the North American academy, on the other hand, the key point is that this trend emerged or is emerging precisely at the time when the political conditions for its reception were/are just right, having originated in and been complicit with the neoliberal university and "educational neoliberalism."[16]

What made Big History compelling (at least for some) was its market success. In 2008, the billionaire and Microsoft cofounder Bill Gates tracked down Christian and offered him financial backing to develop an online course for high school students. By 2019, over fifteen hundred schools around the world used the resulting curriculum. Big History spread from Australia to South Korea, the United States, Canada, the United Kingdom, and the Netherlands.[17] With their reliance on Massive Open Online Courses, the nonjudgmental language of science, and "all matter of data analytics," "big historians" seem to have positioned their movement to appeal to the world of commercialized education, the neoliberal corporate economy, and "Big Data."[18]

It is, perhaps, unsurprising that Big History sells well, in Russia as everywhere in the world. Its uncritical embrace in a postsocialist Russia that wants to forget its socialist past is, however, an ironic finale to this book, which has attempted to rediscover the continuing legacy of the Soviet socialist project in an unexpected place: the historical relationship between the natural sciences and the discipline of history in the twentieth century. Even while the intellectuals whose complex life stories have been traced here had little in common, socialist imaginaries remained the central axis around which their programs of scientific history revolved in the twentieth century. These socialist imaginaries—always changing, never quite realized—continue to inflect scientific history as it is imagined and enacted in the neoliberal university today.

Past Futures of the History of Science

The same year Francis Fukuyama contemplated the end of history, another unorthodox historian, Thomas S. Kuhn, was contemplating the future of the history of science, the field most commonly associated with his groundbreaking 1962 *The Structure of Scientific Revolutions*. In a grant proposal submitted to the National Science Foundation (NSF) at the end of 1988, Kuhn explained that he had taken a year of unpaid leave from the Massachusetts Institute of Technology, where he taught the history and philosophy of science, to work on a book project tentatively titled "Words and Worlds: An Evolutionary View of Scientific Development." Like *The Structure of Scientific Revolutions*, the project was ambitious: it sought to break new ground and offer a radically new vision to future scholars. Kuhn hoped to get the manuscript ready by 1991.[19] He must have thought of the book as the capstone of his career as he planned to retire that year.

The book never materialized, but the proposal gives us a glimpse of what Kuhn had in mind. The project had occupied him since the publication of *The Structure of Scientific Revolutions*, in part because, he explained, that book was misread by so many readers. It is, indeed, notorious for the variety of ways in which it was read and interpreted, all the while thoroughly transforming the way historians and sociologists thought about science.[20] It is now widely regarded as a touchstone for the postpositivist history of science.[21] But this was not what Kuhn had in mind. He now proposed a work that would develop what he saw as the misunderstood conceptual core of his famed work: the connection between biological evolution and scientific history.[22]

In *The Structure of Scientific Revolutions*, Kuhn claimed that scientific revolutions literally change the world. In an often-quoted passage, he wrote: "When paradigms change, the world itself changes with them."[23] Many readers found this claim problematic, and others found it fascinating, but most of Kuhn's interpreters, it seems, misread 'his intentions. The point, he explained in his NSF proposal, comes explicitly only "in the last pages of the original edition of *Structure*: the parallel, that is, between biological and cognitive evolution." The proposal portrayed the human/intellectual and natural/biological realms as deeply entrenched within an evolutionary framework. His readers, as Kuhn put it, "mistook the symptoms of revolutionary change for its substance and sometimes entirely missed the book's central problematic." What he intended to do in this new project was "to work out a coherent philosophical underpinning for the historical or developmental view of scientific development first adumbrated almost thirty years ago."[24]

Kuhn was awarded the NSF grant even though the reviews were lukewarm. "In what sense is he arguing that the development of science is evolutionary? How seriously is the analogy to biological evolution to be taken?" one reviewer asked.[25] Other reviewers omitted comment on the content of Kuhn's proposal altogether, recommending it for funding on the grounds of his credentials as, as one reviewer put it, "clearly the most influential historian and philosopher of the past third of a century." And, after all, as another reviewer put it: "The request for one semester and one summer [of paid leave] is modest for the task he proposes."[26]

Obviously, Kuhn's "biologism" sounded outmoded to his reviewers, most of whom, in different ways, identified with what came to be called *postpositivist* conceptions of science, where *positivist* referred to the logical positivism of the Vienna Circle philosophers.[27] Writing in 1985, the philosopher Arthur Danto commented on the irony that '*The Structure of Scientific Revolutions* was originally commissioned as a contribution to the *International Encyclopedia of Unified Science*, that "monument to positivism." Danto stated: "[Kuhn's] theory of scientific revolutions subverted the enterprise that sponsored it." In *The Structure of Scientific Revolutions*, he explained, "there now really was a unity of science, in the sense that all of science was brought under history rather than, as before, history having been brought under science construed on the model of physics."[28] Had Danto been privy to Kuhn's proposal for "Words and Worlds," he might have found himself wondering whether he was reading the same author. In the NSF proposal, Kuhn sounded as if he wanted to put history under biology.

Somewhat ironically, Kuhn's vision for that project resonated more with nineteenth-century attempts to analogize biological evolution and historical development than with twentieth-century aspirations to use genetics as a resource to overcome the reductionism and historical determinism of earlier accounts of humanity.[29] The reviewers' lukewarm response is understandable. While, in hindsight, his call to bridge biology and history might sound almost prophetic in light of today's search within the history profession for productive ways in which historians can engage with genetics and the neurosciences, in his last project Kuhn seems to have missed a century-long history of attempts to reconcile historical scholarship with contemporary developments in biology, some of which have been mapped out in this book.

Archive Abbreviations

APRF: Archive of the President of the Russian Federation, Moscow.
ARAN: Archive of the Russian Academy of Sciences, Moscow.
GARF: State Archive of Russian Federation (Gosudarstvennyi arkhiv Rossiiskoi Federazii), Moscow.
Garfield Papers: The Papers of Eugene Garfield, The Science Institute Archives, Philadelphia.
IGY Papers: Papers of the International Geophysical Year, Archives of the National Academy of Sciences, Washington, DC.
Julian and Juliette Huxley Papers: Julian and Juliette Huxley Papers, 1899–1988, MS 512, Woodson Research Center, Fondren Library, Rice University, Houston.
Julian Huxley Papers: Julian Sorell Huxley Papers, 1899–1980, MS 50, Woodson Research Center, Fondren Library, Rice University, Houston.
Juliette Huxley Papers: Juliette Huxley Papers 1895–1994, MS 474, Woodson Research Center, Fondren Library, Rice University, Houston.
Kuhn Papers: The Papers of Thomas S. Kuhn, MC.0240, Massachusetts Institute of Technology's Archives and Special Collections, Cambridge, MA.
Needham Papers: Joseph Needham Papers, Cambridge University Library, Cambridge.
RGAE: Russian State Archive of Economy (Rossiiskii gosudarstvennyi arkhiv ekonomiki), Moscow.
RGANI: Russian State Archive of Contemporary History (Rossiiskii gosudarstvennyi arkhiv noveishei istorii), Moscow.
RGASPI: Russian State Archives of Social and Political History (Rossiiskii gosudarstvennyi arkhiv sotsialno-politicheskoi istorii), Moscow.
UNESCO Archives: UNESCO Archives, Paris.

Notes

PREFACE

1. See, e.g., Julia Adeney Thomas, "History and Biology in the Anthropocene: Problems of Scale, Problems of Value," *American Historical Review* 119, no. 5 (2014): 1587–1607; and Daniel Lord Smail, *On Deep History and the Brain* (Berkeley and Los Angeles: University of California Press, 2008). For further discussion, see the introduction below.

2. David Christian, "Bridging the Two Cultures: History, Big History, and Science," *Historically Speaking* 6, no. 5 (2005): 21–26.

3. See "AHR Forum: Historiographic 'Turns' in Critical Perspective," *American Historical Review* 117, no. 3 (2012): 698–813.

4. Thomas, "History and Biology in the Anthropocene," 1603.

INTRODUCTION

1. Arnold J. Toynbee, *Survey of International Affairs, 1931* (London: Royal Institute of International Affairs, 1932), 1.

2. Arnold J. Toynbee, *A Study of History*, vol. 1, *Introduction: The Geneses of Civilizations* (London: Oxford University Press, 1934), 172–77. For further discussion see Ian Hall, " 'Time of Troubles': Arnold J. Toynbee's Twentieth Century," *International Affairs* 90 (2014): 23–36.

3. Ian Hall, "The 'Toynbee Convector': The Rise and Fall of Arnold J. Toynbee's Anti-Imperial Mission to the West," *European Legacy* 17, no. 4 (2012): 455–69, and " 'Time of Troubles.' "

4. While the gendered component of the story is beyond the scope of this book, a good starting point for a consideration of this aspect is "Scientific Masculinities," ed. Erika Lorraine Milam and Robert A. Nye, special issue, *Osiris*, vol. 30, no. 1 (2015).

5. Henri Berr, "Rapport sur l'organisation materielle et la vie scientifique du Centre" (1929), cited in Michel Blay, "Henri Berr et l'histoire des sciences," *Henri Berr et la culture de XXe siècle: Histoire, science et philosophie: Actes du colloque international 24–26 octobre 1994, Paris* (Paris: Albin Michel, 1997), 121–38, 133. For context and details, see chapter 1 below.

6. For an accessible introduction to Ranke, see Antony Grafton, *The Footnote: A Curious History* (Cambridge, MA: Harvard University Press, 1997), 34–61 ("Ranke: A Footnote about Scientific History"). For a general overview of the beginning of professional history, see, e.g., Anna Green and Kathleen Troup, *The Houses of History: A Critical Reader in Twentieth-Century History and Theory* (New York: New York University Press, 1999), esp. chap. 1.

7. For a sampling of the different meanings of *scientific history*, see Peter Novick, *That Noble Dream: The "Objectivity Question" and the American Historical Profession* (Cambridge: Cambridge University Press, 1988); Joyce Appleby, Lynn Hunt, and Margaret Jacob, *Telling the Truth about History* (New York: Norton, 1994); Georg G. Iggers, *Historiography in the Twentieth Century: From Scientific Objectivity to the Postmodern Challenge* (Middletown, CT: Wesleyan University Press, 1997); Green and Troup, *The Houses of History*; Rens Bod, Jaap Maat, and Thijs Weststeijn, eds., *The Making of the Humanities*, vol. 3, *The Modern Humanities* (Amsterdam: Amsterdam University Press, 2014); and Ian Hesketh, *The Science of History in Victorian Britain* (London: Pickering & Chatto, 2011).

8. See, e.g., A. K. Mayer, "Fatal Mutilations: Educationism and the British Background to the 1931 International Congress for the History of Science and Technology," *History of Science* 40, no. 4 (2002): 445–72.

9. Actors' categories, unlike interpretative or analytic categories, imply using a term in the sense it was used and understood by the historical figures being described. Paying special attention to actors' categories has been "the prevalent methodology" in the history of science since the 1970s. See Gowan Dawson and Bernard Lightman, introduction to *Victorian Scientific Naturalism: Community, Identity, Continuity*, ed. Gowan Dawson and Bernard Lightman (Chicago: University of Chicago Press, 2014), 1–26, 2. Anna Mayer's groundbreaking work provides further insights into the far-reaching aspirations of early professional practitioners of the history of science. See Anna-K. Mayer, "Roots of the History of Science in Britain, 1916–1950" (PhD diss., University of Cambridge, 2003). I thank Anna Mayer for sharing her work and for insightful conversations that have been crucial to the conception of this project.

10. The history of science served many different goals and had different intellectual and political uses. The story told in this book is one among many others in the history of the field. For a brief overview, see Lorraine Daston, "History of Science," in *International Encyclopedia of the Social and Behavioral Sciences*, ed. Neil J. Smelser and Paul B. Baltes (London: Pergamon, 2001), 6842–48. For general accounts of the intellectual history of the history of science as a field

with varied cultural politics and intellectual agendas, see John R. R. Christie, "The Development of the Historiography of Science," in *Companion to the History of Modern Science*, ed. R. C. Olby, G. N. Cantor, J. R. R. Christie, and M. J. S. Hodge (London: Routledge, 1990), 5–22; Rachel Laudan, "Histories of the Sciences and Their Uses: A Review to 1913," *History of Science* 31, no. 1 (1993): 1–34; Michael Aaron Dennis, "Historiography of Science: An American Perspective," in *Science in the Twentieth Century*, ed. John Krige and Dominique Pestre (Amsterdam: Harwood, 1997), 1–26; Anna-K. Mayer, "Setting Up a Discipline: Conflicting Agendas of the Cambridge History of Science Committee, 1936–1950," *Studies in History and Philosophy of Science, Part A* 31, no. 4 (2000): 665–89; and Lorraine Daston, "The History of Science as European Self-Portraiture," *European Review* 14, no. 4 (2006): 523–26, and "The Secret History of Science and Modernity: The History of Science and the History of Religion" (paper presented at the conference "'The Engine of Modernity': Construing Science as the Driving Force of History in the Twentieth Century," Columbia University, New York, May 2–3, 2017).

11. Thomas, "History and Biology in the Anthropocene," 1603.

12. Dipesh Chakrabarty, "Anthropocene Time," *History and Theory* 57 (2018): 5–32.

13. Dipesh Chakrabarty, "The Climate of History: Four Theses," *Critical Inquiry* 35, no. 2 (2009): 197–222, 201.

14. See, e.g., Sebouh David Aslanian, Joyce E. Chaplin, Kristin Mann, and Ann McGrath, "AHR Conversation: How Size Matters: The Question of Scale in History," *American Historical Review* 118, no. 5 (2013): 1431–72.

15. For representative publications on Big History and deep history, respectively, see David Christian, *Maps of Time: An Introduction to Big History* (Berkeley and Los Angeles: University of California Press, 2004); and Smail, *On Deep History and the Brain*. For a sympathetic discussion of these approaches by historians who do not necessarily identify with these programs, see, e.g., David C. Krakauer, John Lewis Gaddis, and Kenneth Pomerantz, eds., *History, Big History, and Metahistory* (Santa Fe, NM: Santa Fe Institute, 2017).

16. Marianne Sommer, *History Within: The Science, Culture, and Politics of Bones, Organisms, and Molecules* (Chicago: University of Chicago Press, 2016); Nasser Zakariya, *A Final Story: Science, Myth, and Beginnings* (Chicago: University of Chicago Press, 2017); and Deborah R. Coen, *Climate in Motion: Science, Empire, and the Problem of Scale* (Chicago: University of Chicago Press, 2018). I borrow the term *scientific turn* to characterize the current soul-searching within the humanities from Amanda Rees, "Doing 'Deep Big History': Race, Landscape and the Humanity of H. J. Fleure (1877–1969)," *History of the Human Sciences* 32, no. 1 (2019): 99–120. For other critical interventions of historians of science in today's discussions around Big History, deep history, and biohistory, see, e.g., Marianne Sommer, "'It's a Living History, Told by the Real Survivors of the Times—DNA': Anthropological Genetics in the Tradition of Biology as Applied History," in *Genetics and the Unsettled Past: The Collision of DNA, Race, and History*, ed. Keith Wailoo, Alondra Nelson, and Catherine Lee (New Brunswick, NJ: Rutgers University Press, 2012), 225–46; Nasser Zakariya, "Is History Still a Fraud?" *Historical Studies in the Natural Sciences* 43,

no. 5 (2012): 631–41; Deborah R. Coen, "Big Is a Thing of the Past: Climate Change and Methodology in the History of Ideas," *Journal of the History of Ideas* 77, no. 2 (2016): 305–21; and Rees, "Doing 'Deep Big History.' "

17. See Lorraine Daston, "Science Studies and the History of Science," *Critical Inquiry* 35, no. 4 (2009): 798–813.

18. On the distinction between scientific explanation and historical understanding, see chapter 1 below. For a discussion of historical understanding as a scientific concern, see Anna-K. Mayer, "Setting Up a Discipline, II: British History, 1948," *Studies in History and Philosophy of Science, Part A* 35 (2004): 41–72, esp. 43–55. For a philosophical reflection on historical understanding as a scientific concern, see Henk De Regt, Sabina Leonelli, and Kai Eigner, eds., *Scientific Understanding: Philosophical Perspectives* (Pittsburgh, PA: University of Pittsburgh Press, 2009), esp. 189–209 (Sabina Leonelli, "Understanding in Biology: The Impure Nature of Biological Knowledge").

19. Johanna Bockman introduced the term *liminal spaces* as a descriptor of the places in which new knowledge is produced across/outside dichotomies of power and established systems of classification and hierarchies of knowledge. See Johanna Bockman, *Markets in the Name of Socialism: The Left-Wing Origins of Neoliberalism* (Stanford, CA: Stanford University Press, 2011).

20. An early example is the landmark Lucien Febvre with Lionel Bataillon, *Geographical Introduction to History*, trans. E. G. Mountford and J. H. Paxton (New York: Knopf, 1925).

21. See, e.g., Fa-ti Fan, "Modernity, Region, and Technoscience: One Small Cheer for Asia as Method," *Cultural Sociology* 10, no. 3 (2016): 352–68; and Warwick Anderson, "Asia as Method in Science and Technology Studies," *East Asian Science, Technology and Society* 6, no. 4 (2012): 445–51. On the role of the history of science in fomenting the notion of Western modernity, see Daston, "The History of Science as European Self-Portraiture"; and Marwa Elshakry, "When Science Became Western: Historiographical Reflections," *Isis* 101, no. 1 (2010): 98–109.

22. Fan, "Modernity, Region, and Technoscience," 363.

23. See the discussion in Fan, "Modernity, Region, and Technoscience." On the aquacentric approach to history, which uses the oceans to anchor world history, see David Armitage, Alison Bashford, and Sujit Sivasundaram, eds., *Oceanic Histories* (Cambridge: Cambridge University Press, 2018). I thank Minakshi Menon and participants in the working group "History of Science in Asia: Decolonizing the History of Science" at the meeting of the Consortium for History of Science, Technology and Medicine (Philadelphia 2019) for helpful discussions of this topic.

24. I use *Russia* in the same way scholars refer to *Asia* in discussions of Asia as method, i.e., as a shorthand for Russia's situated, textured, and multiethnic experiences without essentializing the region or the peoples inhabiting it.

25. The analytic approaches of Kapil Raj to circulation of knowledge and of Marwa Elshakry to the translation of knowledge between cultures, in particular, have been helpful to think with while writing this book. Both present a complex picture of diverse contacts and mutual relations involved in the process of the exchange of knowledge, in which the so-called imported knowledge and the locale

itself were being profoundly changed as a result of knowledge transfer. See Kapil Raj, *Relocating Modern Science: Circulation and the Construction of Knowledge in South Asia and Europe, 1650–1900* (New York: Palgrave Macmillan, 2006); and Marwa Elshakry, *Reading Darwin in Arabic, 1860–1950* (Chicago: University of Chicago Press, 2013). For a discussion of Kapil Raj's approach to the mobility of scientific knowledge as a particularly suitable analytic lens with which to study Russian/Soviet human and social sciences , see Susan Gross Solomon, "Circulation of Knowledge and the Russian Locale," *Kritika: Explorations in Russian and Eurasian History* 9 (2008): 9–26.

26. For an intellectual history that examines the place of the idea of a universal West in the Muslim world, see Aydin Cemil, *The Politics of Anti-Westernism in Asia: Visions of World Order in Pan-Islamic and Pan-Asian Thought* (New York: Columbia University Press, 2017). For the discussion of the West as a trope in Russian cultural discourse, see chapter 2. For the discussion of the early postcolonial critique of *West* as a loaded category, see chapter 5. Throughout this book, I use *West* as an actors' category rather than an analytic or a conceptual category.

27. Alexander Etkind, *Internal Colonization: Russia's Imperial Experience* (Cambridge: Polity, 2011). See also Michael David-Fox, Peter Holquist, and Alexander Martin, eds., *Orientalism and Empire in Russia* (Bloomington, IN: Slavica, 2006); and Vera Tolz, *Russia's Own Orient: The Politics of Identity and Oriental Studies in the Late Imperial and Early Soviet Periods* (Oxford: Oxford University Press, 2011).

28. For an accessible introduction to the great debate between the Slavophiles and the Westernizers in imperial Russia, see Geoffrey Hosking, *Russia and the Russians: From Earliest Times to 2001* (London: Penguin, 2001).

29. Mikhail Pokrovskii, "Svoeobrazie russkogo istoricheskogo protsessa i pervaia bukva marksizma," *Krasnaya nov'* (1922), http://www.hrono.ru/statii/2001/pokrov_1bukva.html.

30. N. Bukharin, "Theory and Practice from the Standpoint of Dialectical Materialism," in *Science at the Cross Roads: Papers Presented to the International Congress of the History of Science and Technology* (London: Kniga, 1931; reprint, London: Routledge, 2013), 9–33, 32.

31. For an overview of the trend, see David Armitage, *Foundations of Modern International Thought* (Cambridge: Cambridge University Press, 2012), esp. chap. 1 ("The International Turn in Intellectual History"), 17–32.

32. For a rare exception, see "Social and Human Sciences on Both Sides of the 'Iron Curtain,'" ed. Ivan Boldyrev and Olessia Kirtchik, special issue, *History of the Human Sciences* 29, nos. 4–5 (2016): esp. 3–12 (Ivan Boldyrev and Olessia Kirtchik, "On [Im]permeabilities: Social and Human Sciences on Both Sides of the 'Iron Curtain'"). For an insightful account that uses Russia to anchor the global history of the political far Right, see Marlène Laruelle, *Entangled Far Rights: A Russian-European Intellectual Romance in the Twentieth Century* (Pittsburgh, PA: University of Pittsburgh Press, 2018).

33. Oleg Kharkhordin, "From Priests to Pathfinders: The Fate of the Humanities and Social Sciences in Russia after World War II," *American Historical Review* 120, no. 4 (2015): 1283–98, 1290, 1291, 1295.

34. For a discussion of the "French intellectual revolution" in Russia in the 1990s, see Aleksandr Dmitriev, "Russkie pravila dlia frantsuzskoi teorii: Opyt 1990-kh godov," *Respublika slovesnosti: Frantsiia v mirovoi intellektual'noi kul'ture*, ed. S. N. Zenkin (Moscow: NLO, 2005), 177–88.

35. Kharkhordin, "From Priests to Pathfinders," 1296. On Gurevich popularizing the *Annales* school in the Soviet Union in the 1970s, see Mikhail Krom, "Studying Russia's Past from an Anthropological Perspective: Some Trends of the Last Decade," *European Review of History/Revue européene d'histoire* 11, no. 1 (2004): 69–77, 71.

36. For a similar critique in a different context, see Elshakry, *Reading Darwin in Arabic.*

37. I. Borichevsky, "Naukovedenie kak tochnaia nauka," *Vestnik znaniia* 12 (1926): 778–88, 778.

38. Michael M. J. Fischer, "A Tale of Two Genome Institutes: Qualitative Networks, Charismatic Voice, and R&D Strategies—Juxtaposing GIS Biopolis and BGI," *Science, Technology and Society* 23, no. 2 (2018): 271–88.

CHAPTER ONE

1. J. B. Bury, *An Inaugural Lecture Delivered in the Divinity School, Cambridge, on January 26, 1903* (Cambridge: Cambridge University Press, 1903).

2. "Scientific History," *Punch*, February 2, 1910, 89.

3. Lynn Hunt, "French History in the Last Twenty Years: The Rise and Fall of the *Annales* Paradigm," *Journal of Contemporary History* 21, no. 2 (1986): 209–24, 212. For a comprehensive account of the *Annales* school of historiography, see Peter Burke, *The French Historical Revolution: The Annales School, 1929–2014* (Cambridge: Polity, 2015).

4. Bury, *Inaugural Lecture.*

5. See Lucien Febvre, "De 1892 à 1933: Examen de conscience d'une histoire et d'un historien," *Revue de synthèse historique* 7, no. 2 (1934): 93–106; and Michael Harsgor, "Total History: The *Annales* School," *Journal of Contemporary History* 13, no. 1 (1978): 1–13.

6. Lucien Febvre, *Combats pour l'histoire* (1953; reprint, Paris: A. Colin, 1992), 16, 20. See the discussion in Allan Megill, "Coherence and Incoherence in Historical Studies: From the 'Annales' School to the New Cultural History," *New Literary History* 35, no. 2 (2004): 207–31. For a discussion of collective research practices within the *Annales* school, see chapter 5 below.

7. See Friedrich Stadler, *The Vienna Circle: Studies in the Origins, Development, and Influence of Logical Empiricism* (Vienna: Springer, 2001).

8. Marc Bloch and Lucien Febvre, "A nos lecteurs," *Annales d'histoire économique et sociale* 1, no. 1 (1929): 1–2, 2.

9. See Rudolf Carnap, Hans Hahn, and Otto Neurath, "The Scientific Conception of the World: The Vienna Circle," in *Empiricism and Sociology*, ed. Marie Neurath and Robert S. Cohen (Boston: Reidel, 1973), 299–319, 299, 304 (originally published as *Wissenschaftliche Weltauffassung: Der Wiener Kreis*, ed. Verein Ernst Mach [Vienna: Wolf, 1929]).

10. See Rudolf Carnap, *The Logical Structure of the World*, trans. R. George (Chicago: Open Court Classics, 2003), originally published as *Der logische Aufbau der Welt* (Berlin: Weltkreis, 1928); Carnap, Hahn, and Neurath, "The Scientific Conception of the World." For the discussion of the Vienne Circle's "scientific world-conception" as a social attitude, see Donata Romizi, "The Vienna Circle's 'Scientific World-Conception': Philosophy of Science in the Political Arena," *HOPOS: The Journal of the International Society for the History of Philosophy of Science* 2, no. 2 (2012): 205–42.

11. See, e.g., Otto Neurath, Rudolf Carnap, and Charles F. W. Morris, *Foundations of the Unity of Science: Toward an International Encyclopedia of Unified Science*, 2 vols. (Chicago: University of Chicago Press, 1955–70). On the spread of the unity of science movement and its continuing legacies, see Friedrich Stadler and Juha Manninen, eds., *The Vienna Circle and the Nordic Countries: Networks and Transformations of Logical Empiricism* (Dordrecht: Springer, 2010); Vassiliki Betty Smocovitis, *Unifying Biology: The Evolutionary Synthesis and Evolutionary Biology* (Princeton, NJ: Princeton University Press, 1996); Peter Galison, "The Americanization of Unity," *Daedalus* 127, no. 1 (1998): 45–71; and George Reisch, *How the Cold War Transformed Philosophy of Science: To the Icy Slopes of Logic* (Cambridge: Cambridge University Press, 2005.For a recent reconsideration of the subject that places the Vienna Circle's unity of science program in the larger context of various projects of scientific unification championed within the natural sciences, see Harmke Kamminga and Geert Somsen, eds., *Pursuing the Unity of Science: Ideology and Scientific Practice from the Great War to the Cold War* (London: Routledge, 2016).

12. On Reichenbach and the philosophical distinction between the context of discovery and the context of justification, see Jutta Schickore and Friedrich Steinle, eds., *Revisiting Discovery and Justification* (Dordrecht: Springer, 2006).

13. Wilhelm Dilthey, *Einleitung in die Geisteswissenschaften* (Leipzig: Duncker & Humblot, 1883), translated as *Introduction to the Human Sciences: An Attempt to Lay a Foundation for the Study of Society and History* by Ramon J. Betanzos (Detroit: Wayne State University Press, 1988).

14. In contrast to the word *science*, which in English denotes only the natural sciences, the German *Wissenschaft*, the French *les sciences*, and the Russian *nauka* have the broader meaning of "organized knowledge" and can be applied to both the sciences and the humanities. Even so, these linguistic traditions distinguish between the exact sciences and the interpretative humanities: *Natur-* vs. *Geisteswissenschaft* in German, *les sciences* vs. *les lettres* in French, *естественные* vs. *гуманитарные науки* in Russian, etc. For this reason, when translating *Geisteswissenschaft*, I use the word *humanities* rather than the usual *human science*. The French *lettres* corresponds to what in English is called *the humanities*, while *les sciences humaines* can denote both the human sciences (e.g., psychology) and the humanistic disciplines (e.g., history).

15. The historian Frederick Beiser reconstructed Dilthey's reasoning on the basis of his lecture notes: "To know is to derive from first principles; to explain is to subsume under general mathematical laws; and to understand is to interpret or translate, to make someone's meaning comprehensible to me by putting it in my own terms." Frederick Beiser, *The German Historicist Tradition*

(Oxford: Oxford University Press, 2012), 298, quoted in Lydia Patton, "Methodology of the Sciences," in *The Oxford Handbook of German Philosophy in the Nineteenth Century*, ed. Michael Forster and Kristin Gjesdal (Oxford: Oxford University Press, 2015), 594–606. For further discussion of the matter, see Uljana Feest, ed., *Historical Perspectives on Erklären and Verstehen* (Dordrecht: Springer, 2009).

16. See Eric S. Nelson, "Dilthey and Carnap: The Feeling of Life, the Scientific Worldview, and the Elimination of Metaphysics," in *The Worlds of Positivism: A Global Intellectual History, 1770–1930*, ed. Johannes Feichtinger, Franz L. Fillafer, and Jan Surman (Cham: Palgrave Macmillan, 2018), 321–46. While the members of the Vienna Circle saw their program as a social and political project, they did not develop a "political philosophy of science" and always insisted on separating their scientific and logical work from issues of culture, politics, and history. For the later history of the Vienna Circle in their exile in Cold War America, where the philosophers were compelled to deemphasize the cultural, political, and pedagogical aspects that crucially defined their earlier project, see Reisch, *How the Cold War Transformed Philosophy of Science*. For a succinct historiographic discussion of the "Left Vienna Circle thesis," see Sarah S. Richardson, "The Left Vienna Circle, Pt. 1: Carnap, Neurath, and the Left Vienna Circle Thesis," *Studies in History and Philosophy of Science, Part A* 40, no. 1 (2009): 14–24.

17. Rens Bod, Jaap Maat, and Thijs Weststeijn, "Introduction: The Making of the Modern Humanities," in Bod, Maat, and Weststeijn, eds., *The Modern Humanities*, 13–26.

18. For a critical discussion, see Werner Callebaut, *Taking the Naturalistic Turn; or, How Real Philosophy of Science Is Done* (Chicago: University of Chicago Press, 1993).

19. See Guy Ortolano, *The Two Cultures Controversy: Science, Literature, and Cultural Politics in Postwar Britain* (Cambridge: Cambridge University Press, 2009).

20. For an insider account of the Vienna Circle, see Victor Kraft, *The Vienna Circle: The Origin of Neo-Positivism: A Chapter in the History of Recent Philosophy* (New York: Greenwood, 1953).

21. This point was argued succinctly in Enrico Castelli Gattinara, *Les inquiétudes de la raison: Épistémologie et histoire en France dans l'entre-deux-guerres* (Paris: Vrin/Éditions de l'EHESS, 1998).

22. Quoted in Mary Pickering, *Auguste Comte: An Intellectual Biography*, 3 vols. (Cambridge: Cambridge University Press, 1993–2009), 3:2.

23. Auguste Comte, *Cours de philosophie positive*, 4 vols. (Paris: Rouen, 1830–42). The first English translation was an abbreviated one: Harriet Martineau: *The Positive Philosophy of Auguste Comte* (London: J. Chapman, 1853). For a comprehensive overview of Comte's life and work, see Pickering, *Auguste Comte*.

24. *Cours*, I, L 1, 21, quoted in Warren Schmaus, "Comte's General Philosophy of Science," in *Love, Order, and Progress: The Science, Philosophy, and Politics of Auguste Comte*, ed. Michel Bourdeau, Mary Pickering, and Warren Schmaus (Pittsburgh, PA: University of Pittsburgh Press, 2018), 27–55, 34.

25. Auguste Comte, *Comte: Early Political Writings*, ed. H. S. Jones (Cambridge: Cambridge University Press, 1998), 83. For discussions of different aspects of Comte's work, see Bourdeau, Pickering, and Schmaus, eds., *Love, Order, and Progress*.

26. At least in part, the success might be attributed to the fact that Comte's writings belonged to an emerging and immensely popular genre of science popularization that offered accessible articulations of the meanings and the place of science in modern life at the time of widespread debates regarding the new industrial, machine-governed world. On the role of visionary nonspecialized books on science in this period, see James A. Secord, *Visions of Science: Books and Readers at the Dawn of the Victorian Age* (Chicago: University of Chicago Press, 2014). For a discussion of Comte's synthesis as a narrative genre, see Zakariya, *A Final Story*, esp. 68–72.

27. On Comte and Mill, see Michael Singer, *The Legacy of Positivism* (New York: Palgrave Macmillan, 2005). On Littré's early career as a translator of Hippocrates and a historian of medicine, see Roger Rullière, "Les études médicales d'Émile Littré," *Revue de synthèse historique* 103 (1982): 255–62. For Littré's later career as a politician during the early years of the Third Republic, see Sudhir Hazareesingh, *Intellectual Founders of the Republic: Five Studies in Nineteenth-Century French Political Thought* (New York: Oxford University Press, 2001).

28. See Warren Schmaus, Mary Pickering, and Michel Bourdeau, "The Significance of Auguste Comte," in Bourdeau, Pickering, and Schmaus, eds., *Love, Order, and Progress*, 3–24.

29. Auguste Comte, *Système de politique positive; ou, Traité de sociologie instituant la religion de l'humanité*, 4 vols. (Paris: Carilian-Goeury, 1851–54), translated as *System of Positive Polity* by J. H. Bridges (vol. 1), F. Harrison (vol. 2), E. S. Beesly et al. (vol. 3), R. Congreve (vol. 4), and H. D. Hutton ("Appendix: Early Essays") (London: Longmans, Green, 1875–77), and *Catéchisme positiviste; ou, Sommaire exposition de la religion universelle en treize entretiens systématiques entre une femme et un prêtre de l'humanité* (Paris, 1852), translated as *The Catechism of Positive Religion* by Richard Congreve (London: Trubner, 1891).

30. See Andrew Wernick, "The Religion of Humanity and Positive Morality," in Bourdeau, Pickering, and Schmaus, eds., *Love, Order, and Progress*, 217–49.

31. Denise Phillips, "Trading Epistemological Insults: 'Positive Knowledge' and Natural Science in Germany, 1800–1850," in Feichtinger, Fillafer, and Surman, eds., *The Worlds of Positivism*, 137–54, 148. In an essay on the emergence of quantum mechanics during the Weimar period, the historian of science Paul Forman has argued that a widespread "repudiation of positivist conceptions of the nature of science" has created a favorable climate for new ideas of acausality. According to Forman: "The movement to dispense with causality in physics, which . . . blossomed . . . in Germany after 1918, was primarily an effort by German physicists to adapt the content of their science to the values of their intellectual environment. . . . If the physicist were to improve his public image he had first and foremost to dispense with causality, with rigorous determinism, that most universally abhorred feature of the physical world picture." And it was that

abhorred feature that positivism had come to epitomize in the German academy. Paul Forman, "Weimar Culture, Causality, and Quantum Theory: Adaptation by German Physicists and Mathematicians to a Hostile Environment," *Historical Studies in the Physical Sciences* 3 (1971): 1–115, 7.

32. While Franz Brentano helped spread Comte's ideas in Austria, while a professor at the University of Vienna, Ernst Mach, whose philosophy inspired much of the Vienna Circle's program, made only cursory and largely dismissive remarks about Comte. See Franz L. Fillafer and Johannes Feichtinger, "Habsburg Positivism: The Politics of Positive Knowledge in Imperial and Post-Imperial Austria, 1804–1938," in Feichtinger, Fillafer, and Surman, eds., *The Worlds of Positivism*, 191–238. On Mach's and the Vienna Circle's roots in the cultural and intellectual history of Vienna and the Habsburg Empire, see Deborah R. Coen, *Vienna in the Age of Uncertainty: Science, Liberalism, and Private Life* (Chicago: University of Chicago Press, 2007).

33. See, e.g., W. M. Simon, *European Positivism in the Nineteenth Century: An Essay in Intellectual History* (Ithaca, NY: Cornell University Press, 1963); and, more recently, Isabel Noronha-DiVanna, *Writing History in the Third Republic* (Newcastle: Cambridge Scholars, 2010).

34. For a contemporary description of the "Nouvelle Sorbonne" and its triple faith in science, history, and republicanism, see Jules Rouvier, *L'enseignement a l'Université de Paris* (Paris, 1893). For a survey of the role that science and education played in the political life of the early Third Republic, see Robert Fox, *The Savant and the State: Science and Cultural Politics in Nineteenth-Century France* (Baltimore: Johns Hopkins University Press, 2012). For a discussion of the deeply gendered representations of the social order in the pedagogical programs of the early Third Republic, see Judith Surkis, *Sexing the Citizen: Morality and Masculinity in France, 1870–1920* (Ithaca, NY: Cornell University Press, 2006).

35. Émile Littré, *Principes de philosophie positive* (Paris, 1868), 75. On Littré's role as a quasi-official philosopher of the Third Republic, see Hazareesingh, *Intellectual Founders of the Republic.*

36. Mary Pickering, "The Legacy of Auguste Comte," in Bourdeau, Pickering, and Schmaus, eds., *Love, Order, and Progress*, 250–304, 262–64.

37. On the spread of positivism as an organized movement around the globe that set up various "laboratories of positivism," see Feichtinger, Fillafer, and Surman, eds., *The Worlds of Positivism.*

38. For an overview of French historiography in this period, see Noronha-DiVanna, *Writing History in the Third Republic.*

39. Ceri Crossley, "History as a Principle of Legitimation in France (1820–48)," in *Writing National Histories: Western Europe since 1800*, ed. Stefan Berger, Mark Donovan, and Kevin Passmore (London: Routledge, 1999), 49–56.

40. On the dissemination of the Rankean method of historical training, see, e.g., Antony Grafton, "In Clio's American Atelier," in *Social Knowledge in the Making*, ed. Charles Camic, Neil Gross, and Michèle Lamont (Chicago: University of Chicago Press, 2011), 89–117.

41. For a discussion of the *méthodiques*, see Noronha-DiVanna, *Writing History in the Third Republic.*

42. Numa Denis Fustel de Coulanges, *La monarchie franque* (Paris: Hachette, 1888), 32–33.

43. Charles-Victor Langlois and Charles Seignobos, *Introduction aux études historiques* (Paris: Hachette, 1898), 1–2.

44. Noronha-DiVanna, *Writing History in the Third Republic*, 226.

45. Charles Seignobos, *La méthode historique appliquée aux sciences sociales* (Paris: Félix Alcan, 1901).

46. Noronha-DiVanna, *Writing History in the Third Republic*, 229.

47. Henri Berr, "Sur notre programme," *Revue de synthèse historique* 1 (1900): 1–8.

48. Henri Berr, *La synthèse en histoire: Essai critique et théorique* (Paris: Alcan, 1911), 3, 23.

49. See, e.g., Fritz Ringer, *Fields of Knowledge: French Academic Culture in Comparative Perspective, 1890–1920* (New York: Cambridge University Press, 1992), on 276–77.

50. Lucien Febvre, "Sur une forme d'histoire qui n'est pas la nôtre," *Annales* 3, no. 1 (1948): 21–24. As Dosse noted, the Annalistes built their identity attacking the so-called positivistic history. See François Dosse, *New History in France: The Triumph of the Annales* (Urbana: University of Illinois Press, 1994), chap. 1.

51. Henri Berr, "Essais sur la science de l'histoire: La méthode statistique et la question des grands hommes," *La nouvelle revue* 64 (1890): 517–23, 517–18.

52. The theoretical part of the thesis appeared in print as Henri Berr, *L'avenire de la philosophie: Esquisse d'une synthèse des connaissances fondée sur l'histoire* (Paris: Hachette, 1899). On Berr's consistent focus on these themes, see Enrico Castelli Gattinara, "L'idée de la synthèse: Henri Berr et les crises du savoir dans la première moitié du XXe siècle," in *Henri Berr et la culture de XXe siècle: Histoire, science et philosophie: Actes du colloque international, 24–26 octobre 1994, Paris* (Paris: Albin Michel, 1997), 21–38.

53. Henri Poincaré, "L'évolution des lois," *Scientia* 9 (1911): 275–92. Echoing Poincaré, Berr wrote in his major book-length statement on his philosophy of history, *Traditional History and Historical Synthesis*: "It is possible that there are universal and eternal laws, the permanent nature of things; but the laws, for the most part and without doubt, *are born*, and once they are born they are not irrefutable [*inattaquable*] and do not stay unchangeable." Henri Berr, *L'histoire traditionnelle et la synthèse historique* (Paris: F. Alcan, 1921), 43. For a discussion of the impact of Poincaré's "L'évolution des lois" on young French philosophers of history, see Isabel Gabel, "From Evolutionary Theory to Philosophy of History: Raymond Aron and the Crisis of French Neo-Transformism," *History of the Human Sciences* 31, no. 1 (February 2018): 3–18.

54. Henri Berr, "Au bout de dix ans," *Revue de synthèse historique* 21 (1910): 5, cited in Castelli Gattinara, "L'idée de la synthèse," 35.

55. Berr, "Sur notre programme," 1–2.

56. George Sarton, "Henri Berr (1863–1954): La synthèse de 1'histore et 1'histoire de la science," *Centaurus* 4 (1956): 185–97.

57. Quoted in Martin Siegel, "Henri Berr's *Revue de synthèse historique*," *History and Theory* 9 (1970): 322–34, 328. In 1910, Febvre joined the editorial board of the *Revue de synthèse historique*, in charge of the section "The Regions of France."

58. Henri Berr, preface to *L'évolution de l'humanité: Synthèse collective: Introduction générale* (Paris: Renaissance du livre, 1920), v–xxviii.

59. Lucien Febvre, *La terre et l'évolution humaine: Introduction géographique à l'histoire* (Paris: Renaissance du Livre, 1922).

60. Cristina Chimisso, *Writing the History of the Mind: Philosophy and Science in France, 1900–1960s* (Aldershot: Ashgate, 2008), 89. On the center more generally, see ibid., 87–93.

61. For a list of members of the center's administrative council, see "Section de synthèse historique," *Bulletin du Centre international de synthèse* 8 (1929): 2–3.

62. On the weeks of synthesis, see Bernadette Bensaude-Vincent, "Présence scientifiques aux semaines de synthèse," *Revue de synthèse historique* 117, nos. 1–2 (1996): 219–30; and Marina Neri, "Vers une histoire psychologique: Henri Berr et les semaines internationales de synthèse (1929–1947)," *Revue de synthèse historique* 117, nos. 1–2 (1996): 205–18.

63. Bensaude-Vincent, "Présence scientifiques aux semaines de synthèse."

64. See, e.g., the published proceedings of the first week of synthesis: Maurice Caullery, Émile Guyénot, and P. Rivet, *L'évolution en biologie: Première semaine internationale de synthèse* (Paris: Renaissance du livre, 1929); and Lucien Febvre, Émile Tonnelat, Marcel Mauss, Adfredo Niceforo, and Louis Weber, *Civilisation: Le mot et l'idée: Première semaine international de synthèse* (Paris: Renaissance du livre, 1930). The proceedings of the scientific and the humanities sessions were published separately, reflecting the dual theme of the week, evolution and civilization. For further discussion of the humanities sessions at the first week of synthesis, see chapter 3 below.

65. Quoted in Harry R. Ritter and T. C. R. Horn, "Interdisciplinary History: A Historiographical Review," *History Teacher* 19, no. 3 (1986): 427–48, 435.

66. For the proceedings of the meeting of the International Committee of the History of Science and the First International Congress of the History of Science, see "Comptes rendus de la première session du Comité international d'histoire des sciences et du Premier congrès international d'histoire des sciences (Paris, 20–24 mai 1929)," *Archeion* 11, suppl. (1929): i–cxi.

67. Brigitte Schroeder-Gudehus and Anne Rasmussen, *Les fastes du progrès: Le guide des expositions universelles, 1851–1992* (Paris: Flammarion, 1992).

68. On international scientific congresses, see Robert Fox, *Science without Frontiers: Cosmopolitanism and National Interests in the World of Learning, 1870–1940* (Corvallis: Oregon State University Press, 2016).

69. On the history of the historical congresses, see Karl Dietrich Erdmann, *Toward a Global Community of Historians: The International Historical Congresses and the International Committee of Historical Sciences, 1898–2000* (New York: Berghahn, 2005).

70. See W. Rawley, "L'histoire aux Congrès de 1900," *Revue de synthèse historique* 1, no. 1 (1900): 196–208, 208.

71. "Congrès international d'histoire comparé, Paris 1900: Liste générale des membres," in *Annales internationales d'histoire: Congrès de Paris, 1900*, 7 vols. (Paris: Armand Colin, 1901–2), 1:v–xliv, xi.

72. Comte, *Cours*, II, L 49, 173n, quoted in Schmaus, "Comte's General Philosophy of Science," 27.

73. See Gattinara, *Les inquiétudes de la raison en France dans l'entre-deux-guerres*; and Chimisso, *Writing the History of the Mind.*

74. H. W. Paul, "Scholarship and Ideology: The Chair of the General History of Science at the Collège de France, 1892–1913," *Isis* 67 (1976): 376–97; Annie Petit, "L'héritage du positivisme dans la création de la Chaire d'histoire générale des sciences au Collège de France," *Revue d'histoire des sciences* 48 (1995): 521–56.

75. See *Annales internationales d'histoire*, vol. 5.

76. *Congrès international d'histoire comparée, tenu à Paris du 23 au 28 juillet 1900: Procès-verbaux sommaires* (Paris: Imprimerie nationale, 1901), 15.

77. *Annales internationales d'histoire*, 1:8.

78. See *Congrès international d'histoire comparée, tenu à Paris du 23 au 28 juillet 1900.*

79. On the comparative method in Comte's system, see Laurent Clauzade, "Auguste Comte's Positive Biology," in Bourdeau, Pickering, and Schmaus, eds., *Love, Order, and Progress*, 93–127.

80. Rawley, "L'histoire aux congrès de 1900," 208.

81. *Annales internationales d'histoire*, 5:5–26, 164–70.

82. Gaston Milhaud, "Sur un point de la philosophie scientifique d'Auguste Comte," in *Annales internationales d'histoire*, 5:15–26.

83. E. Gley, "Influence du positivism sur le développement des sciences biologiques en France," in *Annales internationales d'histoire*, 5:164–70, 168, 167.

84. The successful competitor for the chair was Gabriel Monod, one of the *méthodiques* historians Berr criticized.

85. Cited in Michel Blay, "Henri Berr et l'histoire des sciences," in *Henri Berr et la culture de XXe siècle*, 121–38, 133.

86. On Mieli, see Cristina Chimisso, "Fleeing Dictatorship: Socialism, Sexuality and the History of Science in the Life of Aldo Mieli," *History Workshop Journal*, no. 72 (2011): 30–51.

87. Caullery, Guyénot, and Rivet, *L'évolution en biologie*, i.

88. At the time of the first week of synthesis, Caullery was a member of the International Committee of the History of Science. See "Liste des membres du Comité international d'histoire des sciences," *Archeion* 11 (1929): 118–21, 120.

89. George Sarton, *Introduction to the History of Science*, 3 vols. in 5 pts. (Baltimore: Williams & Wilkins, 1927–48), 1:32. Sarton was committed to Comte's teaching throughout his life, even making a pilgrimage to his *domicile sacré* in Paris to "commune" with his spirit. He designed *Isis* to spread positivist teaching. In an article published in the opening issue of *Isis*, he stated that Comte was the "founder of the history of science" and that his own work adhered to Comtean school of thought. George Sarton, "L'histoire de la science," *Isis* 1, no. 1 (1913): 3–46. As Arnold Thackray and Robert K. Merton noted in their discussion of Comte's legacy in institutionalizing the history of science in the United States, both Sarton's arguments and his actions stemmed from a mixture of "positivism, progressivisme, and Utopian socialism." Arnold Thackray and Robert K. Merton, "On Discipline Building: The Paradoxes of George Sarton," *Isis* 63, no. 4 (1972): 473–95, 479. More recently, John F. M. Clark argued that Sarton "built on eighteenth- and nineteenth-century traditions of positivism and

universal history." John F. M. Clark, "Intellectual History and the History of Science," in *A Companion to Intellectual History*, ed. Richard Whatmore and Brian Young (Oxford: Wiley & Sons, 2016), 155–69, 159. On Sarton's life and work, see Lewis Pyenson, *The Passion of George Sarton: A Modern Marriage and Its Discipline* (Philadelphia: American Philosophical Society, 2007), 187, 276, 445–47.

90. On Sarton, see Pyenson, *The Passion of George Sarton.*

91. See, e.g., George Sarton, "The New Humanism," *Isis* 6, no. 1 (1924): 9–42.

92. George Sarton, *The History of Science and the New Humanism* (Cambridge, MA: Harvard University Press, 1937), 3rd ed. (New Brunswick, NJ: Transaction, 1988), 32.

93. George Sarton, "L'histoire de la science et l'organisation internationale," *La vie internationale* 4 (1913): 27–40. See also the discussion of this article in Bert Theunissen, "Unifying Science and Human Culture: The Promotion of the History of Science by George Sarton and Frans Verdoorn," in Kamminga and Somsen, eds., *Pursuing the Unity of Science*, 182–206.

94. Theunissen, "Unifying Science and Human Culture," 188.

95. Geert J. Somsen, "A History of Universalism: Conceptions of the Internationality of Science from the Enlightenment to the Cold War," *Minerva* 46, no. 3 (2008): 361–79, 368–70.

96. Pyenson, *The Passion of George Sarton*, 93.

97. Lewis Pyenson and Christophe Verbruggen, "Ego and the International: The Modernist Circle of George Sarton," *Isis* 100, 1 (2009): 60–78, 77.

98. Hendrick De Man, *Zur Psychologie des Sozialismus* (Jena: E. Diederichs, 1927). On Sarton and De Man, see Christophe Verbruggen and Lewis Pyenson, "History and the History of Science in the Work of Hendrick De Man," *Belgisch Tijdschrift voor Nieuwste Geschiedenis* 41 (2011): 487–511.

99. Daniel Laqua, "Transnational Intellectual Cooperation, the League of Nations, and the Problem of Order," *Journal of Global History* 6, no. 2 (2011): 223–47, 245.

100. See John Culbert Faries, *The Rise of Internationalism* (New York: W. D. Gray, 1915), 12.

101. As, e.g., in Murray's *New English Dictionary* (1901): "International character or spirit: the principle of community of interests or action between different nations." For further examples of the use of the term in this period, see Faries, *The Rise of Internationalism.*

102. See Fox, *Science without Frontiers*; and Martin H. Geyer and Johannes Paulmann, eds., *Science without Frontiers, Society, and Politics from the 1840s to the First World War* (New York: Oxford University Press, 2008).

103. See, e.g., Roger Chickering, *Karl Lamprecht: A German Academic Life (1856–1915)* (Atlantic Highlands, NJ: Humanities, 1993).

104. The largest section, that devoted to historical synthesis, directed jointly by Berr and Febvre, listed thirty French and thirty foreign members. Other sections planned to have ten French and ten foreign members each. *Archeion* 11 (1929): 438–39.

105. Laqua, "Transnational Intellectual Cooperation," 245.

106. Giuliana Gemelli, "Communauté intellectuelle et stratégies institutionnelles: Henri Berr et la fondation du Centre international de synthèse," *Revue de synthèse* 108, 2 (1987): 225–59, 246.

107. "Section de synthèse historique," 2. Besides Luchaire, the center's administrative council included the French minister of public education, Pierre Marraud, and the then senator Paul Doumer.

108. IIIC programs ranged from organizing scientific conferences and revising school textbooks to promoting the Universal Decimal Classification as a bibliographic standard to assist with the task of ordering knowledge. See Laqua, "Transnational Intellectual Cooperation."

109. On the history of the term *intellectuals* and its origins in the Dreyfus affair, see Jeremy Jennings and Anthony Kemp-Walsh, "The Century of the Intellectual: From the Dreyfus Affair to Salman Rushdie," in *Intellectuals in Politics: From the Dreyfus Affair to Salman Rushdie*, ed. Jeremy Jennings and Anthony Kemp-Walsh (London: Routledge, 1997), 1–24; and Stefan Collini, *Absent Minds: Intellectuals in Britain* (Oxford: Oxford University Press, 2006).

110. See Lucien Febvre, *De la "Revue de synthèse" aux "Annales": Lettres à Henri Berr, 1911–1954*, ed. J. Pluet and G. Candar (Paris: Fayard, 1997). *Cartel des gauches* refers to a coalition of socialist, radical-socialist, and republican-socialist parties that formed the government of the Third Republic after achieving victory at the polls in 1924. On the *cartel des gauches* and the Left in France, see Tony Judt, *Marxism and the French Left: Studies in Labour and Politics in France, 1830–1981* (New York: Oxford University Press, 1986).

111. On Berr's political attitudes, see Chimisso, *Writing the History of the Mind.*

112. On Lanvegin's political activism, see Isabelle Gouarné, *L'introduction du Marxisme en France: Philosoviétisme et sciences humaines, 1920–1939* (Rennes: Presses universitaires de Rennes, 2013); and Mary Jo Nye, "Science and Socialism: The Case of Jean Perrin in the Third Republic," *French Historical Studies* 9, no. 1 (1975): 141–69.

113. On Langevin's connections with the League of Nations, see Bensaude-Vincent, "Présence scientifiques aux semaines de synthèse," 223.

CHAPTER TWO

1. On international historical congresses as an international arena for scientific history, see chapter 1 above.

2. As Geoffrey Hosking puts it, in the debates about Russia's past, present, and national identity, " 'the West' became hypostatized in Russian cultural discourse as a single homogeneous complex of concepts and institutions in opposition to which 'Russia' had to be defined." Hosking, *Russia and the Russians*, 275. There is a vast literature on the theme of Russia and the West. For a general introduction, see, e.g., Paul Dukes, *World Order in History: Russia and the West* (London: Routledge, 1995); and Russell Bova, ed., *Russia and Western Civilization: Cultural and Historical Encounters* (Armonk, NY: M. E. Sharpe, 2003). For a historiographic overview of the nineteenth-century debates, see Frances Nethercott, "Russia and the West," in *Europe and Its Others: Essays*

on Interperception and Identity, ed. Paul Gifford and Tessa Hauswedell (Bern: Peter Lang, 2010), 225–44.

3. On the continuing legacies of the Slavophile-Westernizer great dispute, see Laura Engelstein, *Slavophile Empire: Imperial Russia's Illiberal Path* (Ithaca, NY: Cornell University Press, 2009).

4. Quoted in Nethercott, "Russia and the West," 225.

5. See O. A. Kiriash, "Praktika puteshestviia kak sposob formirovaniia predstavlenii o prostranstve russkimi istorikami vtoroi poloviny XIX veka," *Dialog so vremenem* 39 (2012): 183–95.

6. A. Bekasova, "Die Formierung eines kulturellen Milieus: Russische Studenten und ihre Reisen im späten 18. Jahrhundert, " in *Die Welt erfahren: Reisen als kulturelle Begegnung von 1780 bis heute*, ed. Arnd Bauerkämper, Hans Erich Bödeker, and Berhard Struck (Frankfurt: Campus, 2004), 239–64.

7. See, e.g., Nikolai I. Kareev, review of *La synthèse en histoire* (1911) by Henri Berr, *Istoricheskoe Obozrenie* 17 (1912): 274–87. On Berr, see chapter 1 above.

8. See Thomas Nemeth, "Positivism in Late Tsarist Russia: Its Introduction, Penetration, and Diffusion," in Feichtinger, Fillafer, and Surman, eds., *The Worlds of Positivism*, 273–91.

9. N. I. Kareev, "Poniatie nauki i klassifikatsiia nauk," with commentary by A. V. Malinov, *Klio* 2 (2013): 28–35. See also N. I. Kareev, *Vvedenie v izuchenie sotsiologii* (Saint Petersburg: M. M. Stasulevich, 1897).

10. For the discussion of the place of comparison in the Comtean system of positive knowledge, see chapter 1 above.

11. Quoted in Wladimir Berelowitch, "History in Russia Comes of Age," *Kritika: Explorations in Russian and Eurasian History* 9, no. 1 (2008): 113–34, 124.

12. W. Berelowitch, "Istoriia sotsial'naia, natsional'naia, vseobshchaia: 'Zhurnal Ministerstva narodnogo prosveshcheniia' i Istoricheskoe obshchestvo pri Sankt-Peterburgskom universitete na poroge XX veka," in *Istoriograficheskii sbornik* (Saratov: Izdatel'stvo Saratovskgo gosudarstvennogo universiteta, 2010), 43–64.

13. N. I. Kareev, *Krest'iane i krest'ianskii vopros vo Frantsii v poslednei chetverti XVIII veka* (Moscow: Izd-vo M. N. Lavrova, 1879), 218, quoted in Nethercott, "Russia and the West," 234. On the role of the big questions that dominated the nineteenth century, see Holly Case, *The Age of Questions* (Princeton, NJ: Princeton University Press, 2018).

14. N. Kareiew, *Les paysans et la question paysanne en France dans le dernier quart du XVIIIe siècle* (Paris: V. Giard & E. Brière, 1899).

15. Nethercott, "Russia and the West," 238. On Fustel de Coulanges, see chapter 1 above.

16. See Erdmann, *Toward a Global Community of Historians*.

17. For the list of participants, see "Congrès international d'histoire comparé."

18. A. G. Slonimskii, "Uchastie rossiiskikh istorikov v mezhdunarodnykh kongressakh istorikov," *Voprosy istorii* 7 (1970): 95–108.

19. Slonimskii, "Uchastie rossiiskikh istorikov v mezhdunarodnykh kongressakh istorikov," 96.

20. Erdmann, *Toward a Global Community of Historians*, 45.

21. Slonimskii, "Uchastie rossiiskikh istorikov v mezhdunarodnykh kongressakh istorikov."

22. In 1911, a large number of professors, among them Vinogradov, resigned from Moscow University in protest of the dismissal of some of their colleagues (the so-called Kasso affair).

23. Slonimskii, "Uchastie rossiiskikh istorikov v mezhdunarodnykh kongressakh istorikov," 99.

24. Cited in Dukes, *World Order in History*, 99.

25. In his long essay on the philosophical presuppositions of Auguste Comte's sociology, e.g., Lappo-Danilevskii provided a detailed list of the oversimplifications and deficiencies of Comte's social theory. A. S. Lappo-Danilevskii, "Osnovnye printsipy sotsiologicheskoi doktriny O. Konta," in *Problemy idealizma*, ed. P. I. Novgorodtsev (Moscow: Izdanie Moskovskogo Psikhologicheskogo Obshchestva, 1902), 394–490. For a discussion, see Alexander Vucinich, *Social Thought in Tsarist Russia: The Quest for a General Science of Society, 1861–1917* (Chicago: University of Chicago Press, 1976), 110–21.

26. There exists a vast literature on Lappo-Danilevskii, mostly in Russian. For a general overview of his life and work, see A. V. Malinov and S. N. Pogodin, *Aleksandr Lappo-Danilevskii: Istorik i filosof* (Saint Petersburg: Iskustvo, 2001); and E. A. Rostovtsev, *A. S. Lappo-Danilevskii i Peterburgskaia istoricheskaia shkola* (Riazan': NRII, 2004).

27. I. M. Grevs, "Aleksandr Sergeevich Lappo-Danilevskii," *Russkii istoricheskii zhurnal* 6 (1920): 44–81.

28. See T. I. L'vovich, "A. S. Lappo-Danilevskii i arkheologiia," *Klio* 84, no. 12 (2013): 115–19. For a discussion of how the discoveries in archaeology have straddled the boundaries of history, see James Turner, *Philology: The Forgotten Origins of the Modern Humanities* (Princeton, NJ: Princeton University Press, 2014). I thank Lorraine Daston for pointing out this connection to me.

29. Lappo-Danilevskii, "Osnovnye printsipy sotsiologicheskoi doktriny O. Konta."

30. A. S. Lappo-Danilevskii, *Metodologiia istorii* (Saint Petersburg: Lit. Bogdanova, 1909). A substantively reworked and enlarged edition of *Metodologiia istorii* appeared in 1910 and 1913. See A. S. Lappo-Danilevskii, *Metodologiia istorii: Posobie k lektsiiam chitaemym studentam S.-Peterb. un-ta* (Saint Petersburg: Stud. izd. Ist.-filol. fak., 1910); and A. S. Lappo-Danilevskii, *Metodologiia istorii*, vol. 1, *Teoriia istoricheskogo znaniia*, and vol. 2, *Metody istoricheskogo izucheniia* (Saint Petersburg: Stud. izd. Ist.-filol. fak., 1913).

31. Alexander Vucinich, *Science in Russian Culture, 1861–1917* (Stanford, CA: Stanford University Press, 1970), 119.

32. L. N. Mazur, "'Vizual'nyi povorot' v istoricheskoi nauke na rubezhe XX–XXI vv.: V poiskakh novykh metodov issledovaniia," *Dialog so Vremenem* 46 (2014): 95–108.

33. On the place of biology in Comte's system, see chapter 1 above.

34. On Lappo-Danilevskii's assessment of the epistemological role of biology in the study of society, see A. V. Malinov, "Biologicheskoe napravlenie v sotsiologii: Iz lektsii akademika A. S. Lappo-Danilevskogo (opyt rekonstruktsii)," *Peterburgskaia sotsiologiia segodnia* 1 (2014): 217–29.

35. See Vucinich, *Science in Russian Culture*; and Aleksandr Dmitriev, " 'Science nationale' et importations culturelles," *Cahiers du monde russe* 51, no. 4 (2010): 1–29.

36. On the meanings and uses of liberal philosophy in Russia, see, e.g., Vanessa Rampton, *Liberal Ideas in Tsarist Russia: From Catherine the Great to the Russian Revolution* (Cambridge: Cambridge University Press, 2020).

37. Slonimskii, "Uchastie rossiiskikh istorikov v mezhdunarodnykh kongressakh istorikov," 4.

38. Anna Geifman, "The Kadets and Terrorism, 1905–1907," *Jahrbücher für Geschichte Osteuropas* 36, no. 2 (1988): 248–67.

39. Sheila Fitzpatrick, *The Cultural Front: Power and Culture in Revolutionary Russia* (Ithaca, NY: Cornell University Press, 1992).

40. The vision of building a new culture and creating new people had deep roots in the movements that emerged in the wake of the 1860s reforms. See Michael David-Fox, *Revolution of the Mind: Higher Learning among the Bolsheviks, 1918–1929* (Ithaca, NY: Cornell University Press, 1997).

41. See David G. Rowley, *Millenarian Bolshevism, 1900–1920: Empiriomonism, God-Building, Proletarian Culture* (New York: Taylor & Francis, 2017).

42. Quoted in John Barber, *Soviet Historians in Crisis, 1928–1932* (New York: Holmes & Meier, 1981), 23.

43. *Kniga dlia chteniia po istorii srednikh vekov, sostavlennaia kruzhkom prepodavatelei pod redaktsiei prof. P. G. Vinogradova*, 4 vols. (Moscow: Tip. A. I. Mamontova, 1896–99).

44. For Pokrovskii's biography, see A. A. Chernobaev, *"Professor s pikoi," ili tri zhizni istorika M. N. Pokrovskogo* (Moscow: Lit, 1992).

45. M. N. Pokrovskii, *Russkaia istoriia s drevneishikh vremen* (1896–99), in *Pokrovskii: Izbrannye proizvedeniia*, 4 vols. (Moscow: Mysl', 1965–67), 1:207.

46. M. N. Pokrovskii, *Ocherk istorii russkoi kul'tury*, 2 vols. (Moscow: Mir, 1914–18), *Russkaia istoriia v samom szhatom ocherke*, 3 vols. (Moscow: GIZ, 1920–23), and *Ocherki istorii Revoliutsionnogo dvizheniia v Rossii XIX i XX vv.* (Moscow: GIZ, 1924).

47. M. N. Pokrovskii, "Obshchestvennye nauki v SSSR za 10 let (doklad na konferentsii marksistko-leninskikh uchrezhdenii 22 marta 1928 g.)," *Vestnik Kommunisticheskoi Akademii* 26 (1928): 5–6. Ironically, Pokrovskii's phrase "history is nothing but politics projected into the past" is often quoted, pejoratively, as the motto he allegedly endorsed himself.

48. Pokrovskii, *Russkaia istoriia v samom szhatom ocherke*, 3:182. See the discussion in Korine Amacher, "Mikhail N. Pokrovsky and Ukraine: A Normative Marxist between History and Politics," *Ab Imperio* 2018, no. 1 (2018): 101–32.

49. Pokrovskii, *Russkaia istoriia v samom szhatom ocherke*, 3:515.

50. M. N. Pokrovskii, "K istorii SSSR," *Istorik-Marksist* 17 (1930): 18.

51. Amacher, "Mikhail N. Pokrovsky and Ukraine," 102.

52. See Nikolai Krementsov, *Stalinist Science* (Princeton, NJ: Princeton University Press, 1997); and Alexei B. Kojevnikov, *Stalin's Great Science: The Times and Adventures of Soviet Physicists* (London: Imperial College Press, 2004).

53. Lesley Chamberlain, *Lenin's Private War: The Voyage of the Philosophy Steamer and the Exile of the Intelligentsia* (New York: St. Martin's, 2006).

54. See David-Fox, *Revolution of the Mind.*

55. David-Fox, *Revolution of the Mind.*

56. Pokrovskii also helped launch and wrote historical entries for the *Great Soviet Encyclopedia.* On the history of the *Great Soviet Encyclopedia,* see O. E. Bogomazova, "Sovetskie entsyklopedii 1920–1960-kh godov kak istochnik istoriograficheskikh predstavlenii," *Vestnik Cheliabinskogo gosudarstvennogo universiteta* 22 (2014): 161–66; and Brian Kassof, "A Book of Socialism: Stalinist Culture and the First Edition of the *Bol'shaia sovetskaia entsiklopediia,*" *Kritika: Explorations in Russian and Eurasian History* 6 (2005): 55–95.

57. Barber, *Soviet Historians in Crisis,* 19.

58. Not coincidentally, the first postwar international congress of historians met in Belgium, a country that championed the neutrality and independence guaranteed by the London Treaty of 1839. On the special status of Belgium and the notion of neutrality, see Daniel Laqua, *The Age of Internationalism and Belgium, 1880–1930: Peace, Progress and Prestige* (Manchester: Manchester University Press, 2015).

59. Cited in Erdmann, *Toward a Global Community of Historians,* 105.

60. Erdmann, *Toward a Global Community of Historians.*

61. On the Soviet-German weeks of science held in Berlin in the summer of 1928, see Annette Vogt, "Soviet-German Scientific Relations," in *World Views and Scientific Discipline Formation,* ed. W. R. Woodward and R. S. Cohen (Dordrecht: Kluwer Academic, 1991), 391–99.

62. See D. A. Aleksandrov et al., *Sovetsko-Germanskie nauchnye sviazi vremeni Veimarskoi respubliki* (Saint Petersburg: Nauka, 2001).

63. See M. N. Pokrovskii, "Doklad o poezdke v Oslo," *Vestnik Kommunisticheskoi akademii* 30, no. 6 (1928): 231–37.

64. Erdmann, *Toward a Global Community of Historians,* 129 (and chap. 9 generally).

65. For the abstract of Pokrovskii's paper, see *Bulletin of the International Congress of Historical Sciences* 5 (1928): 901.

66. Pokrovskii, "Doklad o poezdke v Oslo," 233, 234.

67. M. N. Pokrovskii, "Klassovaia bor'ba i 'ideologicheskii front,'" *Pravda,* November 7, 1928, reprinted in M. N. Pokrovskii, *Istoricheskaia nauka i bor'ba klassov: Istoriograficheskie ocherki, kriticheskie stat'i i zametki* (Moscow: Librokom, 2012), 2:331. For an interpretation of the Bolshevik elite as members of an apocalyptic millennial cult, see Yuri Slezkine *The House of Government: A Saga of the Russian Revolution* (Princeton, NJ: Princeton University Press, 2017). While Pokrovskii is not mentioned in the book, the study provides a window onto the culture of the close-knit circle of "old Bolsheviks" to which he certainly belonged and a context within which to situate his rhetorical devices.

68. I. V. Stalin, "God velikogo pereloma," *Pravda,* November 7, 1929, 1, and *Sochineniia,* 13 vols. (Moscow, 1946–55), 12:174.

69. "O reforme deiatel'nosti uchenykh uchrezhdenii i shkol vysshykh stupenei v Rossiiskoi Sotsialisticheskoi Federativnoi Sovetskoi Respubliki," *Vestnik narodnogo prosveshcheniia Soiuza kommun Severnoi oblasti*, nos. 6–8 (1918): 69, quoted in E. I. Kolchinskii, *V poiskakh sovetskogo soiuza filosofii i biologii* (Saint Petersburg: Dmitrii Bulanin, 1999), 221.

70. N. I. Bukharin and E. A. Preobrazhensky, *Azbuka kommunizma* (Moscow: GIZ, 1919), translated as *The ABC of Communism: A Popular Explanation of the Program of the Communist Party of Russia* by Eden Paul and Cedar Paul (Ann Arbor: University of Michigan Press, 1966), 240. For discussion, see David-Fox, *Revolution of the Mind*, 46.

71. For a general discussion of the Great Break in higher learning, see David-Fox, *Revolution of the Mind*, 254–72.

72. *Poslednie novosti* (Paris), December 5, 1929, cited in Nadezhda Platonova, *Istoriia arkheologicheskoi mysli v Rossii vtoroi poloviny XIX—pervaia tret' XX veka* (Saint Petersburg, 2010).

73. Quoted in Amacher, "Mikhail N. Pokrovsky and Ukraine," 114.

74. See Kendall E. Bailes, *Technology and Society under Lenin and Stalin: Origins of the Soviet Technical Intelligentsia, 1917–1941* (Princeton, NJ: Princeton University Press, 1978).

75. Pavlov reportedly called this ad hoc deal *terrible*. On Pavlov, see Daniel P. Todes, *Ivan Pavlov: A Russian Life in Science* (New York: Oxford University Press, 2014).

76. N. Pavlenko, "'Akademicheskoe delo': Istoriki pod pritselom OGPU," *Nauka i zhizn'* 11 (1999): 26–31. It was in this context that, as we have seen, Pokrovskii called to end "peaceful coexistence."

77. See V. S. Brachev, *Krestnyi put' russkogo uchenogo: Akademik S. F. Platonov i ego "delo"* (Saint Petersburg: Stomma, 2005), and *Travlia russkikh istorikov* (Saint Petersburg: Algoritm, 2006).

78. S. Neretina, "Paradigmy istoricheskogo soznaniia v Rossii nachala veka," *Filosofskie issledovaniia* 3 (1993): 154–86.

79. Several leading historians died in exile. Many of those arrested in relation to the Academic affair returned in a few years, their cases dismissed, only to be arrested again during the late 1930s Great Purges, a campaign meant to eliminate Stalin's political opponents. See Pavlenko, "'Akademicheskoe delo.'"

80. Stalin himself was keen on offering interpretations of Russian history. Thus, in a speech in 1931, he justified the crash mobilization campaign of the Five-Year Plan by drawing on an image of Russian history that he presented as "the continual beatings [Russia] suffered because of its backwardness": "We are fifty or a hundred years behind the advanced countries. We must cover this distance in ten years. Either we will do it, or we will be crushed." I. V. Stalin, "O zadachakh khoziaistvennikov" (speech delivered at the First All-Union Conference of Workers of Socialist Industry, February 4, 1931), https://www.marxists.org/russkij/stalin/t13/t13_06.htm.

81. Bukharin was executed on March 13, 1938, after the show trials that marked the pinnacle of Stalin's Great Purges. For his biography, see Stephen F Cohen, *Bukharin and the Bolshevik Revolution: A Political Biography, 1888–1938* (New York: Oxford University Press, 1980).

82. For the history of the KIZ and the historical documents, see V. M. Orel and G. I. Smagina, eds., *Komissiia po istorii znanii, 1921–1932 gg. sbornik dokumentov* (Saint Petersburg: Nauka, 2003).

83. They are listed as members in *Archeion* 11 (1929): vii; and *Archeion* 12 (1930): 21. On the International Center of Synthesis, see chapter 1 above.

84. On the academy's history, see Alexander Vucinich, *Empire of Knowledge: The Academy of Sciences of the USSR, 1917–1970* (Berkeley and Los Angeles: University of California Press, 1984).

85. Most of Vernadskii's works in the history of science were published posthumously. See V. I. Vernadskii, *Trudy po vseobshchei istorii nauki* (Moscow: Nauka, 1988), and *O nauke*, 2 vols. (Saint Petersburg: PKhGI, 2002).

86. On Vernadskii's cosmism, see George M. Young, *The Russian Cosmists* (Oxford: Oxford University Press, 2012).

87. Krementsov, *Stalinist Science*, 20.

88. Vernadskii's 1921 proposal for the establishment of the KIZ is reprinted in Vernadskii, *O nauke*, 2:343–46.

89. For Vernadskii's biography, see Kendall E. Bailes, *Science and Russian Culture in an Age of Revolutions: V. I. Vernadsky and His Scientific School, 1863–1945* (Bloomington: Indiana University Press: 1990).

90. On the KEPS, see Krementsov, *Stalinist Science*.

91. V. V. Struve, *Mathematischer Papyrus des Staatlichen Museums der Schönen Künste in Moskau* (Berlin: Springer, 1930). The "Moscow Mathematical Papyrus" was at the time one of the oldest known mathematical documents.

92. The congress president, Charles Singer, devoted part of his opening address to a presentation of Struve's findings. The talk was published as Charles Singer, "The Beginnings of Science," *Nature* 128, no. 3218 (1931): 7–10.

93. During the 1930s and 1940s, Struve applied Marx's theory of the "Asiatic mode of production" to interpret the history of ancient Eastern countries, particularly in his field of study, ancient Egypt. For Marx's notion of the Asiatic mode of production, see Stephen P. Dunn, *The Fall and Rise of the Asiatic Mode of Production* (London: Routledge Revivals, 1982). For the critique of Struve's arguments, see Marian Sawer, *Marxism and the Question of the Asiatic Mode of Production* (The Hague: Nijhoff, 1977). On Struve as a historian, see S. B. Krikh, "Kak ne napisat' glavnyi trud vsei zhizni: Sluchai akademika V. V. Struve," *Mir istorika: Istoriograficheskii sbornik* 10 (2015): 241–73.

94. Bukharin asked Vernadskii to stay on the KIZ's advisory board. On Bukharin and Vernadskii as heads of the KIZ, see Orel and Smagina, eds., *Komissiia po istorii znanii*.

95. The invitation is reprinted in Orel and Smagina, eds., *Komissiia po istorii znanii*, 396–97. Bukharin also learned about the congress from the science journalist James Crowther, who cultivated an extensive network of contacts within the Soviet government and visited the Soviet Union in 1930. See Christopher A. J. Chilvers, "Five Tourniquets and a Ship's Bell: The Special Session at the 1931 Congress," *Centaurus* 57 (2015): 61–95, and "The Dilemma of Seditious Men: The Crowther-Hessen Correspondence in the 1930s," *BJHS* 36, no. 4 (2003): 417–35.

96. The minutes of the meeting are reprinted in Orel and Smagina, eds., *Komissiia po istorii znanii*, 397–401.

97. N. I. Bukharin to V. M. Molotov, April 25, 1931, Protokoly Politburo, RGASPI, fond 17, opis' 3, delo 822, list 6, reproduced in *Akademia nauk v resheniiakh Poliburo TsK RKP(b)-VKP(b)-KPSS, 1922–1952*, (Moscow: POSPEN, 2000), 106–7.

98. *Akademia nauk v resheniiakh Poliburo . . . , 1922–1952*, 107–9.

99. For discussion of this point, Chilvers, "Five Tourniquets and a Ship's Bell."

100. As Bukharin reported in a paragraph Stalin highlighted: "In Cambridge, I made personal acquaintance with the most significant physicists in the world: [Ernest] Rutherford, [Arthur] Eddington, [Francis William] Aston, [William] Thomson. In London—with a group of biologists sympathetic to the Soviet Union: [Julian] Huxley, [Lancelot] Hogben (materialist), [Joseph] Needham, and others." Quoted in V. D. Esakov and P. E. Rubinin, *Kapitsa, Kreml' i nauka*, vol. 1, *Sozdanie instituta fizicheskikh problem, 1934–1938* (Moscow: Nauka, 2003), 33. For the original, see APRF, fond 3, opis' 33, delo 198, list 55.

101. One such scientist was Peter Kapitsa, with whom Bukharin stayed for several days in Cambridge. (The visit was not mentioned in the report to Stalin.) Kapitsa commented on the visit in a letter to his mother on July 27, 1931, referring to Bukharin as "a very likable person." The letter is quoted in Esakov and Rubinin, *Sozdanie instituta fizicheskikh problem*, 33.

102. N. I. Bukharin, *Teoriia istoricheskogo materializma: Populiarnyi uchebnik marksistskoi sotsiologii* (Moscow, 1921), translated as *Historical Materialism: A System of Marxist Sociology* (New York: International, 1925). For a detailed discussion of *Historical Materialism*, see Cohen, *Bukharin and the Bolshevik Revolution*, 107–22.

103. *Historical Materialism* quoted in Cohen, *Bukharin and the Bolshevik Revolution*, 112–13.

104. For Marx's comments regarding Comte and positivism, see Feichtinger, Fillafer, and Surman, eds., *The Worlds of Positivism*, 1–27 (introduction). Soviet Marxist antipositivist diatribes go back to Lenin's 1908 polemical assault *Materialism and Empiriocriticism*. For discussion, see David Bakhurst, "On Lenin's 'Materialism and Empiriocriticism,'" *Studies in East European Thought* 70 (2018): 107–19.

105. On the anti-Bukharin campaign of 1930, see Jochen Hellbeck, "With Hegel to Salvation: Bukharin's Other Trial," *Representations* 107, no. 1 (2009): 56–90.

106. *Kommunisticheskaia revolutsiia* 2 (1930): 20, quoted in Cohen, *Bukharin and the Bolshevik Revolution*, 114.

107. Cohen, *Bukharin and the Bolshevik Revolution*, 114.

108. All quotes from Colin Holmes, "Bukharin in England," *Soviet Studies* 24, no. 1 (1972): 86–90.

109. Bukharin, "Theory and Practice from the Standpoint of Dialectical Materialism," 32.

110. N. I. Bukharin, *Izbrannye trudy: Istoriia i organizatsiia nauki i tekhniki* (Leningrad: Nauka, 1988), 487. Stalin himself had proclaimed the motto in a speech announcing the crash-mobilization campaign designed to meet the targets established by the Five-Year Plan.

111. N. I. Bukharin, "Bor'ba dvukh mirov i zadachi nauki" (paper presented at the special session of the Academy of Sciences of the Soviet Union, June 21–27, 1931, Moscow), published in Bukharin, *Izbrannye trudy*, 27–52, 51.

112. Bukharin, "Bor'ba dvukh mirov i zadachi nauki," 47.

113. Bukharin, "Bor'ba dvukh mirov i zadachi nauki," 47.

114. Berr was listed as a *member d'honneur* in the minutes of the meeting held during the proceedings of the congress in London. See *Archeion* 13 (1931): 340. For the discussion of the first congress, see chapter 1 above.

115. For the announcement of the Warsaw congress, see *Archeion* 13 (1931): 79.

116. N. Lukin, "VII Mezhdunarodnyi istoricheskii congress v Varshave," *Istorik-marksist* 5 (1933): 118–29.

117. The split with Stalin culminated in November 1933, in the first purge, when Bukharin was accused of "right deviationism" and stood Party trial. See the stenographic account in " 'Byla li otkrovennoi ispoved'? Materialy partiinoi chistki N. I. Bukharina v 1933 g.," *Voprosy istorii KPSS* 1, no. 3 (1991): 40–63.

118. One observer reported that the panel attracted a genuine "migration of peoples" and that a great deal of disappointment was generated when it was announced that Bukharin and A. Lunacharskii, who was also on the program, had failed to show up. S. Bednarski, "VII Międzynarodowy Kongres Nauk Historycznych, Warszawa-Krakow 21–29 VIII 1933," *Przegląd powszechny* 200 (1933): 140–47.

119. Even in its truncated version, the panel was considered to be one of the high points of the congress. See Bednarski, "VII Międzynarodowy Kongres Nauk Historycznych."

120. *Marxism and Modern Thought* (Westport, CT: Hyperion, 1935), vii–viii (preface), viii.

121. N. I. Bukharin, "Marx's Teaching and Its Historical Importance," in *Marxism and Modern Thought*, 1–90, 294.

122. A. I. Tumenev, "Bourgeois Historical Science," in *Marxism and Modern Thought*, 235–319, 294.

123. *Science at the Cross Roads: Papers Presented to the International Congress of the History of Science and Technology* (London: Kniga, 1931).

124. Gary Werskey, *The Visible College: A Collective Biography of British Scientists and Socialists of the 1930s* (London: Allen & Unwin, 1978). On discussion of the volume in France, see chapter 3 below.

125. B. Hessen, "The Social and Economic Roots of Newton's Principia," in *Science at the Cross Roads*, 151–212. On Hessen's own context and motivation, see Loren R. Graham, "The Socio-Political Roots of Boris Hessen: Soviet Marxism and the History of Science," *Social Studies of Science* 15, no. 4 (1985): 705–22.

126. For a discussion of and further bibliography on this vast subject, see Gerardo Ienna and Giulia Rispoli, "Boris Hessen at the Crossroads of Science and Ideology: From International Circulation to the Soviet Context," *Society and Politics* 13 (2019): 37–63.

127. Quoted in William McGucken, *Scientists, Society, and the State* (Columbus: Ohio State University Press, 1984), 73.

128. V. I. Vernadskii, *Dnevniki: 1926–1934* (Moscow: Nauka, 2001), 260.

CHAPTER THREE

1. For discussion, see Wailoo, Nelson, and Lee, eds., *Genetics and the Unsettled Past.*

2. Brian Sykes, *The Seven Daughters of Eve* (New York: Norton, 2002), 1.

3. See esp. Keith Wailoo, Alondra Nelson, and Catherine Lee, "Introduction: Genetic Claims and the Unsettled Past," in Wailoo, Nelson, and Lee, eds., *Genetics and the Unsettled Past*, 1–10; and Sommer, " 'It's a Living History, Told by the Real Survivors of the Times—DNA.' "

4. In a letter dated June 11, 1931, Vavilov wrote to the biologist M. O. Shapovalov, his longtime friend, about his planned departure "tomorrow, heading south, to Crimea and the Caucasus, for some four weeks or so." Nikolai Vavilov, *Nauchnoe nasledstvo v pis'makh: Mezhdunarodnaia perepiska*, 6 vols. (Moscow: Nauka, 1980–2003), 3:34. On the Soviet delegation to the London congress, see chapter 2 above.

5. RGASPI, fond 17, opis' 163, delo 882, list 104. Yet it did not preclude him from attending a physiological congress in Rome in 1932. *Akademia nauk v resheniiakh Poliburo . . . , 1922–1952*, 120, 122, 123. Until his death in 1934, Pavlov regularly traveled abroad to participate in physiological congresses. On Pavlov, see Todes, *Ivan Pavlov.*

6. "O vkluchenii v sostav delegatsii na Londonskii congress po istorii nauki i tekhniki akad: Vavilova," June 5, 1931, minutes of the meeting of the Politburo on May 5, 1931, and the resolutions, RGASPI, fond 17, opis' 3, delo 828, list 14, reproduced in *Akademia nauk v resheniiakh Poliburo . . . , 1922–1952*, 109.

7. See the discussion in chapter 2 above.

8. *Akademia nauk v resheniiakh Poliburo . . . , 1922–1952*, 109.

9. N. I. Vavilov, "Report on the Trip to Attend the Second International Congress for the History of Science in London" (1931), reprinted in Orel and Smagina, eds., *Komissiia po istorii znanii*, 442–44, and "Addendum to the Report on the Trip to Attend the Second International Congress for the History of Science in London," March 13, 1932, GARF, fond 7668, opis' 1, delo 427, list 61, reprinted in ibid., 483–84.

10. In his report, Vavilov described the meeting dismissively, as "not very interesting." See Vavilov, "Report on the Trip to Attend the Second International Congress for the History of Science."

11. For the program of the fifth (special) session, see *Archeion* 14, no. 4 (1932): 532. Vavilov's paper was published as N. I. Vavilov, "The Problem of the Origin of the World's Agriculture in the Light of the Latest Investigations," in *Science at the Cross Roads*, 97–106.

12. Vavilov, "The Problem of the Origin of the World's Agriculture," 97–106.

13. For the history of the Fertile Crescent concept, see Courtney Fullilove, *The Profit of the Earth: The Global Seeds of American Agriculture* (Chicago: University of Chicago Press, 2017), 205–6. For a discussion of the political usages of the term *Fertile Crescent*, see Thomas Scheffler, " 'Fertile Crescent,' 'Orient,' 'Middle East': The Changing Mental Maps of Southwest Asia," *European Review of History* 10, no. 2 (2003): 253–72. For a contemporary critique of the concept, see, e.g., Albert T. Clay, "The So-Called Fertile Crescent and Desert Bay," *Journal of the American Oriental Society* 44 (1924): 186–201.

14. James Robinson, Charles A. Beard, and James Henry Breasted, *Outlines of European History*, 2 vols. (Boston: Ginn, 1914). For the publication history of the *Outline*, see Jeffrey Abt, *American Egyptologist: The Life of James Henry Breasted and the Creation of His Oriental Institute* (Chicago: University of Chicago Press, 2011).

15. Quoted in David Wengrow, *What Makes Civilization? The Ancient Near East and the Future of the West* (Oxford: Oxford University Press, 2011), 6.

16. On Sarton, see chapter 1 above.

17. James Henry Breasted, *The Edwin Smith Surgical Papyrus*, 2 vols. (Chicago: University of Chicago Press, 1930).

18. George Sarton, "The Edwin Smith Surgical Papyrus by James Henry Breasted," *Isis* 15, no. 2 (1931): 355–67. The essay "East and West" was included in Sarton, *The History of Science and the New Humanism*, 59–111, Breasted discussed on 72. On the relationship between Sarton and Breasted, see George Sarton, "James Henry Breasted (1865–1935): The Father of American Egyptology," *Isis* 34, no. 4 (1943): 289–91.

19. On Struve, see chapter 2 above.

20. Singer, "The Beginnings of Science," 9.

21. Vavilov, "The Problem of the Origin of the World's Agriculture," 102. One of the Soviet delegates in London, Modest Rubinstein, commented at length on Singer's opening address in his report on the trip, in particular, criticizing the congress's president for his shortsightedness and underscoring the striking difference of the Marxist perspective presented by the Soviet delegation. M. Rubinstein, "Doklad na zasedanii presidiuma komakademii o poezdke sovetskoi delegatsii na II Mezhdunarodnyi congress po istorii nauki i tekhniki v London," reprinted in Orel and Smagina, eds., *Komissiia po istorii znanii*, 427–40.

22. Abyssinia was mentioned as a region occupied by the Italian army in the late nineteenth century. See Robinson, Beard, and Breasted, *Outlines of European History*, 2:319.

23. Vavilov, "The Problem of the Origin of the World's Agriculture," 102.

24. Vavilov, "The Problem of the Origin of the World's Agriculture," 105.

25. N. I. Vavilov, *Tsentry proiskhozhdeniia kul'turnykh rastenii/Studies on the Origin of Cultivated Plants*, 2 pts. (Russian, pp. 1–138; English, pp. 139–248) (Leningrad: Vsesoiuznyi Institut prikladnoi botaniki i novykh kul'tur, 1926), 244.

26. N. I. Vavilov, "O vostochnykh tzentrakh proiskhozhdeniya kulturnykh rastenii," *Novyj Vostok* 6 (1924): 291–305, 304.

27. On the Soviet tradition of human genogeography, see Susanne Bauer, "Virtual Geographies of Belonging: The Case of Soviet and Post-Soviet Human Genetic Diversity Research," *Science, Technology, and Human Values* 39, no. 4 (2014): 511–37; and Mark B. Adams, "From 'Gene Fund' to 'Gene Pool': On the Evolution of Evolutionary Language," *Studies in the History of Biology* 3 (1979): 241–85. See also Michael Flitner, "Genetic Geographies: A Historical Comparison of Agrarian Modernization and Eugenic Thought in Germany, the Soviet Union, and the United States," *Geoforum* 34, no. 2 (2003): 175–85.

28. See Bauer, "Virtual Geographies of Belonging." In the 1940s, taxonomists in Mexico participated in a similar enterprise, using mapping as a way

of understanding the evolution of maize. For the canonical text in English, see E. J. Wellhausen, L. M. Roberts, and E. Hernandez X, *Races of Maize in Mexico: Origins, Characteristics, and Distribution* (Cambridge, MA: Bussey Institution of Harvard University, 1952). I thank Helen Curry for pointing this work out to me.

29. On Vavilov's life, see Peter Pringle, *The Murder of Nikolai Vavilov: The Story of the Persecution of One of the Great Scientists of the Twentieth Century* (London: JR, 2009); and Igor G. Loskutov, *Vavilov and His Institute: A History of the World Collection of Plant Genetic Resources in Russia* (Rome: International Plant Genetics Research Institute, 1999).

30. On "preparation for a professorship" in pre-Revolution Russia, see chapter 2 above.

31. See O. Yu. Elina, *Ot tsarskikh sadov do sovetskikh polei: Istoriia sel'skokhoziaistvennykh opytnykh uchrezhdenii XVIII—20-e gody XX v*, 2 vols. (Moscow: Egmont-Rossiia, 2008), 1:2.

32. On the history of the USDA, see Fullilove, *The Profit of the Earth*.

33. Elina, *Ot tsarskikh sadov do sovetskikh polei*, 2:80. On the history of the bureau, see A. A. Fedotova and N. P. Goncharov, eds., *Buro po prikladnoi botanike v gody pervoi mirovoi voiny: Sbornik dokumentov* (Saint Petersburg: Nestor-Istoriia, 2014).

34. On Vavilov's recognition of the role Bateson played in his career as a teacher and a mentor, see N. I. Vavilov, "Vil'iam Betson (W. Bateson), 1861–1926: Pamiati uchitelia," *Trudy po prikladnoi botanike i selektsii* 15 (1926): 499–511.

35. See Erik L. Peterson, "William Bateson from *Balanoglossus* to *Materials for the Study of Variation*: The Transatlantic Roots of Discontinuity and the (Un)naturalness of Selection," *Journal of the History of Biology* 41, no. 2 (2008): 267–305. On Bateson's experiences in Central Asia, see Jeff Eden, " 'The First That Ever Burst, Etc.': William Bateson in the Steppe, 1886–1887," *Studies in Travel Writing* 17, no. 1 (2013): 1–21. For Bateson's own account of his travels, see William Bateson, *Letters from the Steppe Written in the Years 1886–1887*, ed. with an introduction by Beatrice Bateson (London: Methuen, 1928).

36. William Bateson, *Materials for the Study of Variation with Especial Regard for Discontinuity in the Origin of Species* (London: Macmillan, 1894). See the discussion in Peterson, "William Bateson from *Balanoglossus* to *Materials for the Study of Variation*."

37. See Loskutov, *Vavilov and His Institute*; and Tiago Saraiva, "Breeding Europe: Crop Diversity, Gene Banks, and Commoners," in *Cosmopolitan Commons: Sharing Resources and Risks across Borders*, ed. Nil Disco and Eda Kranakis (Cambridge, MA: MIT Press, 2013), 185–211, 190.

38. N. I. Vavilov, "Zakon gomologicheskikh riadov v nasledstvennoi izmenchivosti" (1920), reprinted in *Sel'skoe i lesnoe khoziaistvo* 1–3 (1921): 84–99.

39. V. D. Esakov, "Soosbshchenie N. M. Tulaikova ob otkrytii N. I. Vavilovym zakona gomologicheskikh riadov," *Voprosy istorii estestvoznaniia i tekhniki* 4 (1981): 111–13. See also Yi. N. Vavilov, ed., *Riadom s N. I. Vavilovym: Sbornik vospominanii* (Moscow: Sovetskaia Rossiia, 1973), 53.

40. N. I. Vavilov, *Zakon gomologicheskikh riadov v nasledstvennoi izmenchivosti* (Leningrad: Nauka, 1987), 17–18, 249.

41. William Bateson and Anna Bateson, "On the Variations in Floral Symmetry of Certain Plants Having Irregular Corollas," *Journal of the Linnean Society (Botany)* 28 (1891): 386–422, quoted in Staffan Müller-Wille and Marsha L. Richmond, "Revisiting the Origin of Genetics," in *Heredity Explored: Between Public Domain and Experimental Science, 1850–1930*, ed. S. Müller-Wille and C. Brandt (Cambridge, MA: MIT Press, 2016), 367–94.

42. William Bateson, *Mendel's Principles of Heredity: A Defense* (Cambridge: Cambridge University Press, 1902), quoted in Garland Allen, "Mendel and Modern Genetics: The Legacy for Today," *Endeavour* 27 (2003): 63–68, 67.

43. N. I. Vavilov, "Ocherednye zadachi sel'skokhoziastvennogo rastenievodstva (rastitel'nye bogatstva zemli I ikh ispol'zovanie)," *Izvestiia gosudarstvennogo instituta opytnoi agronomii* 3, nos. 2–4 (1925): 63–71, 69.

44. N. I. Vavilov, "The Law of Homologous Series," *Journal of Genetics* 12, no. 1 (1922): 47–89, 72.

45. On starvation among academics in this period, see A. N. Eremeeva, "*Nakhodias' po usloviiam vremeni v provintsii . . .": Praktiki vyzhivaniia rossiiskikh uchenykh v gody grazhdanskoi voiny* (Krasnodar: Platonov, 2017). See also Todes, *Ivan Pavlov*.

46. For a reconstruction of Vavilov's first presentation outlining the law, see V. D. Esakov, "Zakon gomologicheskikh riadov," *Izvestiia TSKhA* 4 (2012): 71–76.

47. N. I. Vavilov, "Iz genetiki zlakov (K voprosu o faktorakh formoobrazovaniia)," cited in Esakov, "Zakon gomologicheskikh riadov."

48. On Mendeleev's use of the same term, see Michael D. Gordin, *A Well-Ordered Thing: Sharing Resources and Risks across Borders* (New York: Basic, 2004).

49. "Entretiens avec André-Georges Haudricourt," in *Languages and Linguists: Aims, Perspectives, and Duties of Linguistics/Les langues et les linguistes: Buts, perspectives et devoirs de la linguistique*, ed. P. Swiggers (Leuven: Peeters, 1997), 7.

50. N. I. Vavilov, "Ob'iasnitel'naia zapiska v Gosplan ob ekspeditsii Otdela prikladnoi botaniki i selektsii v Afganistan i Severnyi Kitai," ARAN, fond 632, delo 7, listy 2–3.

51. Vavilov, "Ocherednye zadachi sel'skokhoziastvennogo rastenievodstva," 67.

52. See Bernard Debarbieux and Gilles Rudaz, *The Mountain: A Political History from the Enlightenment to the Present*, trans. Jane Marie Todd (Chicago: University of Chicago Press, 2015).

53. Peter Simon Pallas, *Observations sur la formation des montagnes et sur les changements arrivés au globe, particulièrement à l'empire de Russie* (Saint Petersburg: Acta Academiae Scientiarum Imperialis Petropolitanae, 1777).

54. Alexander von Humboldt, *Asie Centrale: Recherches sur les chaines de montagnes et la climatologie comparée*, 3 vols. (Paris: Gide, 1843).

55. See Debarbieux and Rudaz, *The Mountain*.

56. The geographers Debarbieux and Rudaz have proposed the term *political mountains* to highlight the status of mountains as objects of politics from the

mid-nineteenth century through the present. See Debarbieux and Rudaz, *The Mountain.*

57. Bateson, *Letters from the Steppe.* See also Eden, " 'The First That Ever Burst.' "

58. Humboldt, *Asie Centrale.*

59. On the history of the conceptualization of Eurasia in Russian historiography, see Mark Bassin, Sergey Glebov, and Marlène Laruelle, eds., *Between Europe and Asia: The Origins, Theories, and Legacies of Russian Eurasianism* (Pittsburgh, PA: University of Pittsburgh Press, 2015).

60. On the entanglement of Russia's imperial interests in the region and studies by Russian naturalists in the nineteenth century, see A. V. Postnikov, *Skhvatka na "Kryshe Mira": Politiki, razvedchiki i geografy v bor'be za Pamir v XIX veke (monografiya v dokumentakh)* (Moscow: Pamyatniki istoricheskoy mysli, 2001).

61. Halford John Mackinder, "The Geographical Pivot of History," *Geographical Journal* 23, no. 4 (1904): 421–37.

62. Halford John Mackinder, *The World War and After: A Concise Narrative and Some Tentative Ideas* (London: G. Philip & Son, 1924).

63. Alexandre Andreyev, *Soviet Russia and Tibet: The Debacle of Secret Diplomacy, 1918–1930s* (Leiden: Brill, 2003), 80 (referencing K. M. Troianovskii, "Vostok i revolutsiia" [1918]).

64. The most prominent of these routes were the two avenues through Kabul and Kandahar. The third passage, via the Pamirs, which led right into the valley of Kashmir, was very difficult, but it was precisely this route of narrow mountainous trails that the adventurous Bolshevik and Comintern agents would use to sneak into India in the early 1920s. It was, perhaps, the best option given that the political situation in Tehran was highly unfavorable for the Bolsheviks. When the first Bolshevik emissary was sent to Tehran in 1917, he was killed on arrival. The diplomats on the next mission (1918–19) were similarly murdered.

65. N. I. Vavilov, "Afganistanskaia ekspeditsiia: iz doklada na otkrytom zasedanii Soveta GIOA," *Izvestiia GIOA* 3, no. 1 (1925): 82–90.

66. Loskutov, *Vavilov and His Institute.*

67. Pringle, *The Murder of Nikolai Vavilov.*

68. See, e.g., N. I. Vavilov, "Abissinskii dnevnik," n.d., N. I. Vavilov Cabinet-Museum, All-Russian Institute of Plant Genetic Resources, Saint Petersburg. Vavilov's travel diaries were published posthumously as N. I. Vavilov and A. N. Krasnov, *Piat' kontinentov* (Moscow: Mysl', 1987), translated as *Five Continents* by Doris Löve (Rome: IPGRI; Saint Petersburg: VIR, 1997).

69. Clipping of a review by L. Demskii, from *Sel'skokhoziaistvennaia gazeta,* October 31, 1929, in Vavilov Papers, RGAE, fond 1056, opis' 1, delo 164. See also N. I. Vavilov and D. D. Bukinich, *Zemledel'cheskii Afganistan* (Agrarian Afghanistan) (Leningrad: Vsesoiuznyi Institute Prikladnoi Botaniki i Novykh Kul'tur, 1929).

70. 'Clipping of a review by Demskii.

71. Alexandre Andreyev, *The Myth of the Masters Revived: The Occult Lives of Nicholas and Elena Roerich* (Leiden: Brill, 2014), 246.

72. The overview of the exhibit was published as *The Book of America's Making Exposition* (New York: City and State Departments of Education,

1921). The theme of the Russian pavilion was "arts and agriculture." Roerich painted the backdrop, while Vavilov helped with seeds and plants displayed.

73. Borodin moved to the United States after the Revolution and worked at the USDA before Vavilov hired him to manage the New York office of the Bureau of Applied Botany. In their private correspondence, Dobzhansky and Filipchenko characterized him quite negatively as a person and as a specialist. See E. I. Kolchinskii, K. V. Manoilenko, and M. B. Konashev, eds., *U istokov akademicheskoi genetiki v Sankt-Peterburge* (Saint Petersburg: Nauka, 2002). On the New York office of the bureau, which functioned until 1924, see E. B. Truskinov, *Russkoe sel'skokhoziaistvennoe predstavitel'stvo v Amerike v svete perepiski N. I. Vavilova i D. N. Borodina* (Saint Petersburg: VIR, 2012). Borodin was involved in both Roerich's grand Asiatic project and the Bolsheviks' plans for the unification of Asia. According to Andreyev, he probably was a Comintern agent, too. On Borodin's relationship with the Roerichs, see Andreyev, *The Myth of the Masters Revived.*

74. Vavilov's letter to the Dalai Lama quoted in Andreyev, *Soviet Russia and Tibet*, 332. The deal apparently fell apart. On the planned Bolshevik mission to Tibet, see A. I. Andreyev and T. I. Yusupova, "Istoriia odnogo ne sovsem obychnogo puteshestviia: Mongolo-Tibetskaia ekspeditsiia P.K. Kozlova (1923–1926 gg.)," *Voprosy istorii estestvoznaniia i tekhniki* 2 (2001): 51–74.

75. See E. I. Kolchinsky, "Nikolai Vavilov in the Years of Stalin's 'Revolution from Above' (1929–1932)," *Centaurus* 56 (2014): 330–58.

76. Kirill Rossiianov, "Travelling with Bolsheviks: Field Work, Expeditions, and Their Patrons" (paper presented at the annual meeting of the History of Science Society, Pittsburgh, PA, 1999). On Gorbunov and his patronage of science in the 1920s, see K. O. Rossiianov and A. N. Gorbunov, "N. P. Gorbunov i organizatsiia sovetskoi nauki (Interv'yu K. O. Rossianova s A. N. Gorbunovym)," *Voprosy istorii estestvoznaniia i tekhniki* 3 (2004): 89–102; and E. P. Podvigina, "Nikolai Petrovitch Gorbunov," in *Nikolai Petrovitch Gorbunov: Vospominaniia, stat'i, dokumenty* (Moscow: Nauka, 1986), 5–41.

77. For contemporary accounts of the Soviet-German expedition to the Pamirs in 1928, see *Pamirskaia ekspeditsiia 1928 g. Trudy ekspeditsii* (Leningrad: Izd-vo Akad. Nauk SSSR, 1930); W. Rickmer Rickmers, "Die Altai-Pamir-Expedition 1928," *Zeitschrift des Deutschen und Österreichischen Alpenvereins* 60 (1929): 59–63; and Richard Finsterwalder, "Das Expeditionsgebiet im Pamir: Zu den Bild- und Kartenbeilagen der Altai-Pamir-Expedition 1928," *Zeitschrift des Deutschen und Österreichischen Alpenvereins* 60 (1929): 143–60.

78. Simply storing the seeds was not enough; they had to be planted and harvested regularly if the collections' vitality was to be maintained. Also, in order to develop high-yield varieties, Vavilov ran a program of trials planting the seeds over the widest possible geographic range. On the labor necessary to maintain the seed collections, see Helen A. Curry, "From Working Collections to the World Germplasm Project: Agricultural Modernization and Genetic Conservation at the Rockefeller Foundation," *History and Philosophy of the Life Sciences*, vol. 39 (2017), https://doi.org/10.1007/s40656-017-0131-8.

79. See Loskutov, *Vavilov and His Institute.*

80. See the discussion in Saraiva, "Breeding Europe."

81. A copy of Bateson's travel diary during his stay in Leningrad in 1925 is kept in Vavilov Papers See RGAE, fond 1056, opis' 1, delo 412, listy 6 and 13.

82. Vavilov, "O vostochnykh tzentrakh proiskhozhdeniya kulturnykh rastenii," "Tzentry proiskhozhdeniya kulturnykh rastenii," *Trudy po prikladnoi botanike i selektsii* 16, no. 2 (1926): 139–248, "Linneevskii vid kak sistema," *Trudy po prikladnoi botanike, genetike, i selektsii* 26, no. 3 (1931): 109–34, "The Process of Evolution in Cultivated Plants," in *Proceedings of the Sixth International Congress of Genetics* (Menasha, WI: Brooklyn Botanic Garden, 1932), 1:331–42, and "Fitogeograficheskie osnovy selektsii rastenii," in *Teoreticheskie osnovy selektsii rastenii* (Moscow and Leningrad: Sel'khozgiz, 1935), 1:17–75.

83. In the early 1930s, Vavilov attached greatest importance to the Chinese East Asian center, placing the Southern Asiatic center in second place. By the late 1930s, he had reversed the order, arguing that 33 percent of all cultivated plants in the world were to be found in the Southeast Asian center, the second most important center being East Asia, with 20 percent of world's cultivated plant species found there, followed by Southwest Asia, with 14 percent. It was this third center that Vavilov thought of as particularly important for Russia since the composition of the crops traditionally cultivated on the territory of Russia reflected Asian influence.

84. See Saraiva, "Breeding Europe"; and Uge Hosfeld, "Zakhvat nemetskimi fashistami gennogo materiala Vavilovskikh institutov v 1943 g.," in *Na perelome*, ed. E. I. Kolchinsky and M. B. Konashev (Saint Petersburg: SPbIIETRAN, 1999), 2:244–59.

85. On Vavilov's reputation among the plant breeders in Germany, see Saraiva, "Breeding Europe."

86. N. I. Vavilov to F. G. Dobrzhanskii, June 11, 1931, Vavilov Papers, RGAE, fond 1056, opis' 1, delo 272.

87. Theodosius Dobzhansky, "N. I. Vavilov: A Martyr of Genetics, 1887–1942," *Journal of Heredity* 38 (1947): 227–32.

88. For a discussion of dialectical materialism as a genuine philosophy rather than simply an ideology, see Loren R. Graham, *Science and Philosophy in the Soviet Union* (New York: Knopf, 1972).

89. Vavilov and Krasnov, *Piat' kontinentov*, 15.

90. N. I. Vavilov to F. G. Dobrzhanskii, June 11, 1931.

91. A. S. Serebrovskii, "Genogeografiia i genofond sel'skokhoziaistvennykh zhivotnykh SSSR," *Nauchnoe Slovo* 9 (1928): 3–22, 9. On Serebrovskii and the notion of the *genofond* (gene pool), see Adams, "From 'Gene Fund' to 'Gene Pool.'" See also Flitner, "Genetic Geographies."

92. See Bauer, "Virtual Geographies of Belonging."

93. Serebrovskii, "Genogeografiia i genofond sel'skokhoziaistvennykh zhivotnykh SSSR," 12.

94. On Augustin Pyramus de Candolle and the early maps of the geographic distribution of crops, see Nils Güttler, "Drawing the Line: Mapping Cultivated Plants and Seeing Nature in Nineteenth-Century Plant Geography," in *New Perspectives on the History of Life Sciences and Agriculture*, ed. Denise Phillips and Sharon Kingsland (New York: Springer, 2015), 27–51.

95. Vavilov, *Tsentry proiskhozhdeniia kul'turnykh rastenii/Studies on the Origin of Cultivated Plants*, 139 (citing Alphonse de Candolle, *L'origine des plantes cultivée* [Paris: Librairie Germer Baillière, 1882]), 145. Vavilov also cited Alphonse de Candolle's *Géographie botanique raisonnée* (Paris: Masson, 1855) and *La phytographie* (Paris: Masson, 1880).

96. For a biography of de Candolle, see S. R. Mikulinskii, L. A. Markova, and B. A. Starostin, *Al'fons Dekandol' 1806–1893* (Moscow: Nauka, 1973).

97. Vavilov, *Tsentry proiskhozhdeniia kul'turnykh rastenii/Studies on the Origin of Cultivated Plants*, 143.

98. Vavilov, "The Process of Evolution in Cultivated Plants."

99. Vavilov, *Tsentry proiskhozhdeniia kul'turnykh rastenii/Studies on the Origin of Cultivated Plants*, 244–45.

100. J. B. S. Haldane, "Is History a Fraud?" *Harper's Magazine* 161, no. 9 (1930): 470–78, 478. For a discussion of Haldane's essay in light of current reconfigurations within academic history, see Zakariya, "Is History Still a Fraud?" As Jenny Bangham observed, Haldane was also one of the first biologists in Britain who used Vavilov's maps as a reference point to articulate his own claims about history and human difference. See Jenny Bangham, *Blood Relations: Transfusion and the Making of Human Genetics* (Chicago: University of Chicago Press, 2020). I thank Jenny for sharing her unpublished manuscript with me.

101. Haldane, "Is History a Fraud?" 471.

102. See, e.g., Febvre, *De la "Revue de synthèse" aux "Annales."*

103. Quoted in Gouarné, *L'introduction du Marxisme en France*, 113.

104. Charles Parain, "L'antiquité et la conception matérialiste de l'histoire," in *A la lumiere du Marxisme: Essais: Sciences physico-mathematiques, sciences naturelles, sciences humaines*, by Jean Baby et al. (Paris: Editions sociales internationales, 1935).

105. See, e.g., Charles Parain, "L'agriculture de l'ancienne Égypte," *Revue des études sémitiques* 2 (1934): 1–12, and "L'origine des plantes cultivées," *Annales d'histoire économique et sociale* 7, no. 36 (1935): 624–28.

106. Lucien Febvre, "Technique agricole de l'ancienne Égypte," *Annales d'histoire économique et sociale* 8, no. 39 (1936): 296.

107. See Febvre, Tonnelat, Mauss, Niceforo, and Weber, *Civilisation*.

108. Marcel Mauss, "Les civilisations: Éléments et formes," in Febvre, Tonnelat, Mauss, Niceforo, and Weber, *Civilisation*, 82–104, 14, 5. See also Oswald Spengler, *Der Untergang des Abendlandes*, 2 vols. (Vienna: Braumüller, 1918), translated as *The Decline of the West*, vol. 1, *Form and Actuality*, and vol. 2, *Perspectives of World History* (London: George Allen & Unwin Ltd., 1918–22).

109. Staffan Müller-Wille, "Claude Lévi-Strauss on Race, History, and Genetics," *Biosocieties* 5, no. 3 (2010): 330–47.

110. Lévi-Strauss wrote a book-length analysis of Mauss's work portraying him as a harbinger and precursor of the structuralist turn in the humanities. See Claude Lévi-Strauss, *Introduction à l'oeuvre de Marcel Mauss* (Paris: Presses universitaires de France, 1950).

111. See Gouarné, *L'introduction du Marxisme en France*, 202–5.

112. André-Georges Haudricourt, "Correspondance Haudricourt/Mauss (1934–1935), présentée, éditée et annotée par Jean-François Bert," *Le portique*, vol. 27 (2011), http://journals.openedition.org/leportique/2557. See also "Entretiens avec André-Georges Haudricourt." On Haudricourt's interactions with Vavilov, see André-Georges Haudricourt and Louis Hédin, *L'homme et les plantes cultivées* (Paris: Gallimard, 1943).

113. André-Georges Haudricourt, *La technologie, science humaine: Recherches d'histoire et d'ethnologie des techniques* (Paris: Les editions de la Maison des Sciences de l'Homme, 1988), 314. See also André-Georges Haudricourt and Pascal Dibie, *Les pieds sur terre* (Paris: Editions Métailié, 1987), 54–76 ("Le retour d'URSS").

114. For the conceptualization of geographic space as fundamentally social, see Henri Lefebvre, *La production de l'espace* (Paris: Éditions Anthropos, 1974), translated as *The Production of Space* by Donald Nicholson-Smith (Chicago: Blackwell, 1991).

115. Fernand Braudel, *The Mediterranean and the Mediterranean World in the Age of Philip II* (Berkeley and Los Angeles: University of California Press, 1995), 1:25.

116. See Kolchinsky, "Nikolai Vavilov in the Years of Stalin's 'Revolution from Above,'" esp. 335–36.

117. Kolchinsky, "Nikolai Vavilov in the Years of Stalin's 'Revolution from Above,'" 338.

118. Among those arrested was Vavilov's brother-in-law, the agronomist Nikolai Makarov. See V. M. Bautin et al., "Nikolai Pavlovitch Makarov," *Izvestiia TSKhA* 3 (2007): 138–43. The archival footage of the Industrial Party show trial, which was filmed for a propaganda production, has been reconstructed by Sergei Loznitsa. See *The Trial*, dir. Sergei Loznitsa (The Hague: Atoms & Void, 2018).

119. The resulting dossier, Operational File 006854, would eventually grow to contain 136 files and form the basis of Vavilov's arrest in 1940. See Pringle, *The Murder of Nikolai Vavilov*.

120. Gorbunov was arrested by NKVD in 1938 and executed few months later.

121. G. V. Grigor'ev, "K voprosu o tsentrakh proiskhozhdeniia kul'turnykh rastenii," *Izvestiia GAIMK*, vol. 13, no. 9 (1932).

122. On migration theories in archaeology, anthropology, and sociology, see William Y. Adams, Dennis P. Van Gerven, and Richard S. Levy, "The Retreat from Migrationism," *Annual Review of Anthropology* 7 (1978): 483–532; and Peter van Dommelen, "Moving On: Archaeological Perspectives on Mobility and Migration," *World Archaeology* 46 (2014): 477–83.

123. A. A. Formozov, "Arkheologiia i ideologiia," *Voprosy filosofii* 2 (1993): 70–82.

124. One announcement of the Second International Congress of the History of Science and Technology mentioned that the organizers expected three representatives of the GAIMK to participate in the congress. See "Second International Congress of the History of Science and Technology," *Nature* 127, no. 3214 (June 6, 1931): 873–74.

125. S. S. Alymov, "Ekspeditsiia v pervobytnost': Ob odnoi nerealizovannoi mechte sovetskoi etnografii," *Etnograficheskoe obozrenie* 4 (2013): 88–94.

126. On the history of the House of Academicians, see V. Naumov, ed., *Dom akademikov: Istoriia i sud'by* (Saint Petersburg: Sokhranennaia kul'tura, 2016).

127. Vernadskii too lived in the house, and, since many other KIZ members lived there as well, KIZ meetings would often take place in Vernadskii's apartment.

128. V. M. Alpatov, *Istoriia odnogo mifa: Marr i marrism* (Moscow: URSS, 2004), 31.

129. See Michiel Leezenberg, "Soviet Orientalism and Subaltern Linguistics: The Rise and Fall of Marr's Japhetic Theory," in Bod, Maat, and Weststeign, eds., *The Modern Humanities*, 97–112; and Ilya Gerasimov, Sergey Glebov, and Marina Mogilner, "Hybridity: Marrism and the Problems of Language of the Imperial Situation," *Ab Imperio* 1 (2016): 27–68. See also chapter 5 below.

130. Vavilov and Krasnov, *Piat' kontinentov*, 214. For discussion, see Gerasimov, Glebov, and Mogilner, "Hybridity." As Gerasimov et al. pointed out, Stalin's attack on Marr and Stalin's hatred of genetics were part of the same desire to "rehabilitate the cult of 'pure forms' (and thus historicism, nationalism, and racism connected to it)." Ibid., 55.

131. One such archaeologist who turned geologist was A. S. Fedorovsky. See Formozov, "Arkheologiia i ideologiia." A turn from archaeology to geology was facilitated by the interdisciplinary nature of the GAIMK institutes and its practices as well as by the institutional connections between archaeologists and geologists in the Soviet Union at the time. Thus, the GAIMK's Institute of Archaeological Technology (Institut arkheologicheskoi tekhnologii), founded in 1930, was headed by a leading Soviet geologist, Aleksandr Fersman, Vernadskii's disciple and closest colleague.

132. The following discussion of Grigor'ev is based on A. A. Formozov, *Rasskazy of uchenykh* (Kursk: KGMU, 2004), 74–81.

133. Quoted in Formozov, *Rasskazy of uchenykh*, 77.

134. For further discussion of how the Soviet antifascist campaign of the 1930s targeted genetics, see chapter 4 below.

135. Vavilov did not get enough time to show the practical results he had promised. Historians argue, however, that the payoff of Vavilov's efforts came later, when new varieties of resistant plants were developed using his collections. See, e.g., Mark B. Tauger, "Pavel Pateleimonovich Luk'ianenko and the Origins of the Soviet Green Revolution," in *The Lysenko Controversy as a Global Phenomenon* (2 vols.), ed. William deJong-Lambert and Nikolai Krementsov (Cham: Springer International, 2017), 97–127. In a different context, Kaori Iida has argued that, in Japan, geneticists who visited Vavilov's institute assimilated his approach to the "science of breeding." In Japan, Vavilov's interdisciplinary and practical approach was juxtaposed by these scientists to a "discipline of genetics" that grew increasingly isolated from other disciplines (as occurred in American universities). See Kaori Iida, "Genetics and 'Breeding as a Science': Kihara Hitoshi and the Development of Genetics in Japan in the First Half of the Twentieth Century," in Phillips and Kingsland, eds., *New Perspectives on the History of Life Sciences and Agriculture*, 439–58.

136. Vavilov's interrogation files are accessible at http://istmat.info/node/35136. For a harrowing discussion of the Vavilov case and his interrogation, see Vadim J. Birstein, *The Perversion of Knowledge* (Boulder, CO: Westview, 2001).

137. The powerful trope *martyr of genetics* was coined by none other than Dobrzhanskii, whom Vavilov tried so hard to convince to return to the Soviet Union. See Dobzhansky, "N. I. Vavilov: A Martyr of Genetics."

CHAPTER FOUR

1. Julian Huxley to Director General of UNESCO, Memorandum, Paris, January 17, 1949, Julian Huxley Papers, box 118, folder 3.

2. Joseph Mayer and George Sarton, "Notes and Correspondence," *Isis* 16, no. 1 (1931): 124–31, 127. On Singer and the planning of the 1931 London congress, see chapter 2 above.

3. For a good starting point for different aspects of Huxley's scientific life and his social thought, see Kenneth C. Waters and Albert Van Helden, eds., *Julian Huxley: Biologist and Statesman of Science* (Houston: Rice University Press, 1993). For more recent studies that place his scientific endeavors in a larger context, see Sommer, *History Within*; R. S. Deese, *We Are Amphibians: Julian and Aldous Huxley on the Future of Our Species* (Berkeley and Los Angeles: University of California Press, 2015); and Paul T. Phillips, *Contesting the Moral Ground: Popular Moralists in Mid-Twentieth-Century Britain* (Montreal: McGill-Queen's University Press, 2013).

4. Julian Huxley, *Memories* (London: Harper & Row, 1970), 1:5.

5. Huxley, *Memories*, 1:75. See also Julian Huxley, "Some Phenomena of Regeneration in Sycon; with a Note on the Structure of Its Collar-Cells," *Philosophical Transactions of the Royal Society of London: Series B* 202 (1912): 165–89.

6. Julian Huxley, *The Individual in the Animal Kingdom* (Cambridge: Cambridge University Press, 1912). For a discussion of this book and the formative role of Bergson's philosophy on Huxley early in his career, see Emily Herring, " 'Great is Darwin and Bergson his poet': Julian Huxley's Other Evolutionary Synthesis," *Annals of Science* 75, no. 1 (2018): 40–54.

7. J. Morell, *Science at Oxford, 1914–1939: Transforming an Arts University* (Oxford: Clarendon, 1997). On Huxley at Oxford, see Peder Anker, *Imperial Ecology: Environmental Order in the British Empire, 1895–1945* (Cambridge, MA: Harvard University Press, 2001).

8. For a detailed discussion, see Sommer, *History Within*.

9. Krishna R. Dronamraju, *If I Am to Be Remembered: The Life and Work of Julian Huxley with Selected Correspondence* (River Edge, NJ: World Scientific, 1993), 41.

10. See Julian Huxley, *Essays of a Biologist* (New York: Knopf, 1923), *Essays in Popular Science* (London: Chatto & Windus, 1926), and *Religion without Revelation* (London: E. Benn, 1927).

11. See the discussion in Phillips, *Contesting the Moral Ground*, esp. chap. 2.

12. Joseph Needham, ed., *Religion without Revelation* (New York: Macmillan, 1925).

13. Joseph Needham, *The Great Amphibium: Four Lectures on the Position of Religion in a World Dominated by Science* (London: Student Christian Movement Press, 1931), 132.

14. See Gregory Blue, "Joseph Needham—a Publication History," *Chinese Science* 14 (1997): 90–132. For a general discussion, see Zakariya, *A Final Story*.

15. Sarton, *The History of Science and the New Humanism*, xi, 50. See also George Sarton, "The Faith of the Humanist," *Isis* 3, no. 1 (1920): 3–6. While he did not endorse Comte's religion of humanity explicitly, he spoke at the Positivist Church of Humanity in London at least at one occasion, in 1919. See the discussion of Sarton's views on religion in James C. Ungureanu, *Science, Religion and the Protestant Tradition: Retracing the Origins of Conflict* (Pittsburgh, PA: University of Pittsburgh Press, 2019).

16. On Singer at Oxford, see E. A. Underwood, "Charles Singer: A Biographical Note," in *Science, Medicine and History: Essays in the Evolution of Scientific Thought and Medical Practice, Written in Honour of Charles Singer*, ed. E. A. Underwood (Oxford: Oxford University Press, 1953), 1:v–ix.

17. Huxley's contribution appeared in the published proceedings of the summer school. See Julian Huxley, "Science and Religion," in *Science and Civilization* (Oxford Unity series), ed. F. S. Marvin (Oxford: Oxford University Press, 1923), 279–329.

18. Charles Singer, "Scientific Humanism," *The Realist: A Journal of Scientific Humanism* 1, no. 1 (1929): 12–18, 18. On the history of the journal, see Robert Bud, " 'The spark gap is mightier than the pen': The Promotion of an Ideology of Science in the Early 1930s," *Journal of Political Ideologies* 22, no. 2 (2017): 169–81.

19. Bud, " 'The spark gap is mightier than the pen,' " 175.

20. On science writing and its history, see Dorothy Nelkin, *Selling Science: How the Press Covers Science and Technology* (New York: W. H. Freeman, 1995). On the emergence of science journalism as a profession in Britain, see Peter J. Bowler, *Science for All: The Popularization of Science in Early Twentieth-Century Britain* (Chicago: University of Chicago Press, 2009).

21. See the discussion in Bowler, *Science for All*.

22. H. G. Wells, *The Outline of History: Being a Plain History of Life and Mankind*, 2 vols. (New York: Macmillan, 1920). For publication history and discussion, see William T. Ross, *H. G. Wells's World Reborn: The Outline of History and Its Companions* (London: Associated University Presses, 2002); and M. Coren, *Invisible Man: The Life and Liberties of H. G. Wells* (London: Bloomsbury, 1993).

23. See Bowler, *Science for All*.

24. Aldous Huxley to Julian Huxley, March 25, 1927, Juliette Huxley Papers, box 11, folder 4. See also Bowler, *Science for All*, 224–25.

25. H. G. Wells, Julian S. Huxley, and G. P. Wells, *The Science of Life*, 31 pts. (London: Amalgamated Press, 1929–30). On the publication history of *The Science of Life*, see Bowler, *Science for All*, 104–7.

26. At the end of 1930, *The Spectator*, a popular magazine with a wide circulation, placed Huxley above Ernst Rutherford, Bertrand Russell, and James Jeans in its "list of Britain's five best brains." See Roland W. Clark, *The Huxleys* (New York: McGraw-Hill, 1968), 204.

27. Huxley's daily pocket calendar for the summer 1931, Julian Huxley Papers, box 49, folder 1.

28. On Needham's symposium, see Erik L. Peterson, *The Life Organic. The Theoretical Biology Club and the Roots of Epigenetics* (Pittsburgh, PA: University of Pittsburgh Press, 2016), 81–92.

29. Bowler, *Science for All.*

30. J. G. Crowther, *Science in Soviet Russia* (London: Williams & Norgate, 1930).

31. On Crowther's Soviet connections, see Chilvers, "The Dilemmas of Seditious Men," and "Five Tourniquets and a Ship's Bell."

32. Folder "Russia, Intourist Trip, 1931," box 144, folder 7; and box file "Russia, 1931," Julian Huxley Papers, box 176, folders 1–4.

33. Huxley's pocket calendar, summer 1931, Julian Huxley Papers, box 49, folder 1.

34. On PEP and the planning debate in Britain, see Daniel Ritschel, *The Politics of Planning: The Debate on Economic Planning in Britain in the 1930s* (New York: Clarendon, 1997). On planned capitalism, see Bockman, *Markets in the Name of Socialism.*

35. Aldous Huxley eventually withdrew from the trip, citing the need to finish the novel he was writing at the time, *Brave New World.* Aldous Huxley to Julian Huxley, May 5, 1931, Juliette Huxley Papers, box 11, folder 5. After Julian returned from his trip to the Soviet Union, Aldous wrote him: "I have finished my book, thank heaven, almost a week ago. It [would be] hopelessly delayed if I'd gone to Russia." Aldous Huxley to Julian Huxley, August 31, 1931, Juliette Huxley Papers, box 11, folder 5.

36. See Michael David-Fox, *Showcasing the Great Experiment: Cultural Diplomacy and the Soviet Union, 1921–1941* (Oxford: Oxford University Press, 2012); Paul Hollander, *Political Pilgrims: Travel of Western Intellectuals to the Soviet Union, China, and Cuba, 1928–1978* (Oxford: Oxford University Press, 1981); and David C. Engerman, *Modernization from the Other Shore: American Intellectuals and the Romance of Russian Development* (Cambridge, MA: Harvard University Press, 2003).

37. The influential travel dispatches from the Soviet Union included John Reed's *Ten Days That Shook the World* (New York: Boni & Liveright, 1919), H. G. Wells's *Russia in the Shadows* (New York: George H. Doran, 1921), André Gide's *Retour de l'U.R.S.S.* (Paris: Gallimard, 1936), and Lion Feuchtwanger's *Moskau, 1937* (Amsterdam: Querido, 1937). On Soviet political travelogues as a subgenre of travel writing, see Jacques Derrida, "Back from Moscow, in the USSR," in *Politics, Theory and Contemporary Culture*, ed. Mark Poster (New York: Columbia University Press, 1993), 197–235.

38. In 1921, *Publisher's Weekly* listed *Russia in the Shadows* as one of the six top best-selling books of the year, while *The Outline of History* was mentioned only among "books being talked about." See "English Book-Trade News," *Pub-*

lisher's Weekly, February 12, 1921; and "Books Being Talked About," *Publisher's Weekly*, January 15, 1921.

39. Huxley, *Memories*, 1:199. In his diary of the trip, Huxley did not specify who the other members of the group were. "Notes, 1930–1931," Julian Huxley Papers, box 63, folder 2.

40. Huxley's pocket calendar, Julian Huxley Papers, box 49, folder 1.

41. Julian Huxley, *A Scientist among the Soviets* (London: Chatto & Windus, 1932), 60–61, 67.

42. Julian Huxley to Nikolai Vavilov, June 26, 1931, in Vavilov, *Mezhdunarodnaia perepiska*, 2:171.

43. "Notes, 1930–1931."

44. "Notes, 1930–1931."

45. Huxley, *A Scientist among the Soviets*, 34–35, 100.

46. Huxley, *A Scientist among the Soviets*, 101–2.

47. Among the material in Huxley's archive, there is a poster issued by Intourist depicting a postcard sent from a trip to the Soviet Union. The text on the postcard (in French) reads: "This is more than a pleasant journey—this is a journey to the new world."

48. As Martin Willis has put it: "The observations of the travel writer are . . . a blend of the seen and the imagined, . . . a combination of preferred viewing drawn from experience and a perceived ideal view drawn from the imagination." Martin Willis, *Vision, Science and Literature, 1870–1920* (London: Pickering & Chatto, 2011), 119. On travel narratives as a literary genre, see Patrick Holland and Graham Huggan, *Tourists with Typewriters: Critical Reflections on Contemporary Travel Writing* (Ann Arbor: University of Michigan Press, 1998), 8–9; and Paul Fussell, *Abroad: British Literary Traveling between the Wars* (Oxford: Oxford University Press, 1980).

49. Huxley, *A Scientist among the Soviets*, 131.

50. The first Russian translation of Wells's *Russia in the Shadows* was published as Gerbert Uells, *Rossiia vo mgle* (Kharkiv: Gos. Izd-vo Ukrainy, 1922). *Russia in the Shadows* stayed in print throughout the Soviet era, being reissued in 1958, 1964, and 1970.

51. Arthur Koestler, e.g., came to the Soviet Union in 1932 and was given an advance to write a travel narrative. The final manuscript was, however, apparently rejected by the publisher. See David Cesarani, *Arthur Koestler: The Homeless Mind* (New York: Free Press, 1999).

52. See, e.g., R. L. Duffus, "As a Scientist Sees Soviet Russia," *New York Times*, August 21, 1932.

53. What emerged from these efforts was a vision of a new elite of professional planners who would staff the National Planning Commission, a body that would mediate between government and industry and between private and public interests and help determine levels of national income, production, investment, consumption, and employment in different sectors of national economy. See Ritschel, *The Politics of Planning*.

54. Julian Huxley, *TVA: Adventure in Planning* (London: Architectural Press, 1943). *Adventure in Planning* helped establish a positive view of large dam projects and became influential throughout the world after World War II. For

the history of the TVA, see Philip Selznick, *TVA and the Grassroots: A Study in the Sociology of Formal Organization* (Berkeley: University of California Press, 1953); Erwin Hargrove, *Prisoners of the Myth: The Leadership of the Tennessee Valley Authority, 1933–1990* (Princeton, NJ: Princeton University Press, 1993); Ronald C. Tobey, *Technology as Freedom: The New Deal and the Electrical Modernization of the American Home* (Berkeley and Los Angeles: University of California Press, 1996); and David Ekbladh, *The Great American Mission: Modernization and the Construction of an American World Order* (Princeton, NJ: Princeton University Press, 2010).

55. See, e.g., Julian Huxley, *If I Were Dictator* (New York: Harper Bros., 1934).

56. Comparisons of Russia and America date back to the nineteenth century. See, e.g., Alan M. Ball, *Imagining America: Influence and Images in Twentieth-Century Russia* (Lanham: Rowman & Littlefield, 2003). For the contemporary accounts comparing, favorably, the United States and the Soviet Union, see William T. R. Fox, *The Super-Powers: The United States, Britain, and the Soviet Union—Their Responsibility for Peace* (New York: Harcourt, Brace, 1944); Merle Elliott Tracy, *Our Country, Our People, and Theirs* (New York: Macmillan, 1938); and Edmund Wilson, *Travels in Two Democracies* (New York: Harcourt, Brace, 1936). For discussion, see Engerman, *Modernization from the Other Shore.*

57. Arthur Koestler, *Darkness at Noon*, trans. Daphne Hardy (London: Jonathan Cape, 1940). See also Arthur Koestler, *Darkness at Noon*, trans. Philip Boehm (New York: Scribner, 2019). This new translation was based on the original manuscript of *Darkness at Noon*, written in German, which was deemed lost but surfaced in 2015.

58. Quoted from Michael Scammell, *Koestler: The Literary and Political Odyssey of a Twentieth-Century Skeptic* (New York: Random House, 2010), 198.

59. *Great Terror* was coined by the British historian Robert Conquest in his account of the Stalinist years, which he compared to the worst years of the French Revolution. See Robert Conquest, *The Great Terror: Stalin's Purge of the Thirties* (New York: Macmillan, 1968).

60. J. B. S. Haldane, "The Place of Science in Western Civilization," *The Realist* 2, no. 2 (1929): 149–64.

61. See Robert E. Kohler, *Lords of the Fly: Drosophila Genetics and the Experimental Life* (Chicago: University of Chicago Press, 1994).

62. Earlier in 1933, Vavilov had visited the Kaiser Wilhelm Institute for Brain Research in Berlin, where Muller was working. By that time, Muller was already seriously contemplating relocating to the Soviet Union and was practicing his Russian (already quite good, according to his son, for whom he had hired a tutor in Russian). See David E. Muller, oral history interview by Milla Pollock, Cold Spring Harbor, July 15, 2004, http://library.cshl.edu/oralhistory/interview/scientific-experience/becoming-scientist/learning-russian/. Muller accepted Vavilov's offer after Nazi storm troopers attacked several researchers at the Kaiser Wilhelm Institute. Thankful, he expressed his admiration for Soviet science in numerous speeches given throughout the country.

63. See Nathaniel Comfort, *The Science of Human Perfection: How Genes Became the Heart of American Medicine* (New Haven, CT: Yale University Press, 2012).

64. Quoted in Peter J. Kuznick, *Beyond the Laboratory: Scientists as Political Activists in 1930s America* (Chicago: University of Chicago Press, 1987), 134.

65. H. J. Muller, *Out of the Night: A Biologist's View of the Future* (New York: Vanguard, 1935). On Muller's eugenic ideas, see William deJong-Lambert, "H. J. Muller and J. B. S. Haldane: Eugenics and Lysenkoism," in deJong-Lambert and Krementsov, eds., *The Lysenko Controversy as a Global Phenomenon*, 2:103–35.

66. On Muller's letter to Stalin, see Diane B. Paul, " 'Our Load of Mutations' Revisited," *Journal of the History of Biology* 20, no. 3 (1987): 321–35, 325–26.

67. Hermann Muller to Julian Huxley, March 9, 1937, quoted in Nikolai Krementsov, *International Science between the World Wars: The Case of Genetics* (London: Routledge, 2005), 93.

68. The point is detailed in Krementsov, *International Science between the World Wars.*

69. The complicated story of the failed attempt to organize the 1937 genetics congress in Moscow is thoroughly examined in Krementsov, *International Science between the World Wars.*

70. Hermann Muller to Julian Huxley, March 9, 1937, quoted in Krementsov, *International Science between the World Wars*, 90. See also deJong-Lambert, "H. J. Muller and J. B. S. Haldane."

71. See Krementsov, *International Science between the World Wars.*

72. John Langdon-Davis, *Russia Puts the Clock Back: A Study of Soviet Science and Some British Scientists* (London: Victor Gollancz, 1949).

73. Sir Henry Dale, "Protest to USSR," *The Times*, November 6, 1948.

74. See Julian Huxley, *Democracy Marches* (New York: Harper & Bros., 1941), and *Evolution: The Modern Synthesis* (London: G. Allen & Unwin, 1942).

75. For a contemporary account of the event, see W. A. Wooster, "A Scientist Visits Moscow," memorandum to the Association of Scientific Workers, London, 1945, Julian Huxley Papers, box 103, folder 1.

76. Raissa Berg, interview by the author, Paris, 2002. On Berg, see Elena Aronova, "Raissa L'Vovna Berg," Jewish Women's Archive, 2005, http://jwa.org/encyclopedia/article/berg-raissa-lvovna.

77. Huxley's notes in folder "Russia, 1945," Julian Huxley Papers, box 176, folder 2.

78. Huxley's notes in folder "Russia, 1945," Julian Huxley Papers, box 176, folder 2.

79. Huxley had the impression that Lysenko's lecture was arranged at his and Ashby's request.

80. "Summary of Lecture by Lysenko Given at the Request of Julian Huxley and Eric Ashby in the Biological Institute of the Academy of Sciences on 1st June, 1945," Julian Huxley Papers, box 176, folder 1.

81. "Summary of Lecture by Lysenko." Both Berg and Huxley recalled this exchange in their memoirs. See Raisa Berg, *Acquired Traits: Memoirs of a Geneticist from the Soviet Union* (New York: Penguin, 1988); and Huxley, *Memories*, 1:281, 284.

82. Huxley, *Memories*, 1:281, 284. See also Berg, *Acquired Traits.*

83. Huxley's notes in folder "Russia, 1945," Julian Huxley Papers, box 176, folder 2; Raissa Berg, interview by the author.

84. Huxley kept the Intourist menu featuring game cutlets and other delicacies as a souvenir, along with the program of the celebrations. See folder "Russia, 1945," Julian Huxley Papers, box 176, folder 2.

85. Julian Huxley, "Science in the USSR: Evolutionary Biology and Related Subjects," *Nature* 156 (1945): 254–56. See also "Draft Notes of an Informal Meeting Held at the Society for Visiting Scientists on August 28th 1945, to Hear Reports from Some of the British Scientific Delegation Which Had Recently Visited the Soviet Union for the 220th Anniversary of the Russian Academy of Sciences," Julian Huxley Papers, box 176, folder 2.

86. See Nikolai Krementsov, "A 'Second Front' in Soviet Genetics: The International Dimension of the Lysenko Controversy, 1944–1947," *Journal of the History of Biology* 29, no. 2 (1996): 229–50, and *International Science between the World Wars*. See also Michael D. Gordin, "How Lysenkoism Became Pseudoscience: Dobzhansky to Velikovsky," *Journal of the History of Biology* 45, no. 3 (2012): 443–68.

87. Huxley was first appointed a secretary of the preparatory committee drafting a constitution for UNESCO. On his own account, the reasoning for the appointment was that "there was no other suitable Englishman free of commitments." Julian Huxley, "UNESCO Notes" (1946), p. 1, Julian Huxley Papers, box 66, folder 2. When the initial candidate for the position, Alfred Zimmern, a leading British educator and a former deputy director of the Institute for Intellectual Cooperation, became ill and it became clear that the serious surgery required would leave him incapable of taking up the position for some time, Huxley was offered the job. See W. H. G. Armytage, "The First Director-General of UNESCO," in *Evolutionary Studies: Studies in Biology, Economy and Society*, ed. M. Keynes and G. A. Harrison (London: Palgrave Macmillan, 1989), 186–93.

88. "X" (George F. Kennan), "The Sources of Soviet Conduct," *Foreign Affairs* 25, no. 3 (1947): 566–82.

89. The speech was published as Andrey Zhdanov, "O mezhdunarodnom polozhenii" (On the international situation), *Pravda*, October 22, 1947.

90. "Lysenko Crushes Genetics in Russia; Gets Party Backing for His Theories," *New York Times*, August 19, 1948, 5; " 'True' Science," *Wall Street Journal*, August 20, 1948, 4; "The Mind of the Kremlin," *Washington Post*, August 22, 1948, B4; "Marxism as Applied to Growing Tomatoes," *Los Angeles Times*, August 25, 1948, A4; "The Phony Tomato," *Washington Post*, August 26, 1948, 10.

91. Cited in Katarzyna Murawska-Muthesius, "Modernism between Peace and Freedom: Picasso and Others at the Congress of Intellectuals in Wrocław, 1948," in *Cold War Modern: Design, 1945–1970*, ed. David Crowley and Jane Pavitt (London: V&A Publishing, 2008), 33–41.

92. On the broader Cominform-backed peace movement, see Robbie Lieberman, *The Strangest Dream: Communism, Anti-Communism, and the U.S. Peace Movement, 1945–1963* (Syracuse, NY: Syracuse University Press, 2000).

93. Julian Huxley, "Intellectuals at Wroclaw," *The Spectator* 181 (September 10, 1948): 326–27.

94. See the material in Juliette Huxley Papers, box 23, folder 13.

95. I. Zavitch, "Professor Khaksli nedovolen," *Literaturnaia gazeta* 4 (September 7, 1948): 4. On the "struggle for peace in all the world" ("bor'ba za mir vo vsem mire") as a central feature of Soviet foreign policy rhetoric, see Timothy Johnston, "Peace or Pacifism? The Soviet 'Struggle for Peace in All the World,' 1948–54," *Slavonic and East European Review* 86, no. 2 (2008): 259–82.

96. D. Zaslavskii, "S kem oni, mastera kul'tury?" *Pravda*, September 6, 1948, 4. On Zaslavskii, see Elena Aronova, "Russian and the Making of World Languages during the Cold War," *Isis* 108, no. 3 (2017): 643–50.

97. See Julian Huxley, "Soviet Genetics: The Real Issue," *Nature* 4155 (June 18, 1949): 935–42.

98. Julian Huxley, *Heredity, East and West: Lysenko and World Science* (New York: Schuman, 1949), vii. (*Heredity, East and West* was released in the United Kingdom as Julian Huxley, *Soviet Genetics and World Science: Lysenko and the Meaning of Heredity* [London: Chatto & Windus, 1949].)

99. Huxley, *Heredity, East and West*, ii. Huxley's interpretation was part of a broader body of Western work/propaganda depicting the crisis of Lysenkoism as a stand-in for the situation of science in the Soviet Union. See Audra J. Wolfe, *Freedom's Laboratory: The Cold War Struggle for the Soul of Science* (Baltimore: Johns Hopkins University Press, 2018).

100. Huxley, *Heredity, East and West*, esp. chap. 5 ("The Totalitarian Regimentation of Thought").

101. Arthur Koestler, Richard Wright, Louis Fischer, Ignazio Silone, Andre Gide, and Stephen Spender, *The God That Failed: A Confession* (New York: Harper & Bros., 1949).

102. See Hannah Arendt, *The Origins of Totalitarianism* (New York: Harcourt, Brace, 1951).

103. On the history of the concept, see Enzo Traverso, ed., *Le totalitarisme: Le XXe siècle en débat* (Paris: Éditions du Seuil, 2001).

104. For the publication that made *the end of ideology* a commonplace, see Daniel Bell, *The End of Ideology: On the Exhaustion of Political Ideas in the Fifties* (Glencoe: Free Press, 1960). For a critical discussion, see Giles Scott-Smith, *The Politics of Apolitical Culture: The Congress for Cultural Freedom, the CIA and Post-War American Hegemony* (London: Routledge, 2002); and Elena Aronova, "The Congress for Cultural Freedom, *Minerva*, and the Quest for Instituting 'Science Studies' in the Age of Cold War," *Minerva* 50, no. 3 (2012): 307–37.

105. For Julian Huxley's correspondence with Aldous Huxley, see G. Smith, ed., *Letters of Aldous Huxley* (London: Chatto & Windus, 1969), 358–59. See also Ia. M. Gall, *Dzhulian Sorel Khaksli* (Saint Petersburg: Nauka, 2004), 44.

106. Julian Huxley, *UNESCO: Its Purpose and Its Philosophy* (Washington, DC: Public Affairs Press, 1948), 11, 55.

107. See Th. Dobzhansky, *Genetics and the Origin of Species* (New York: Columbia University Press, 1937). For Huxley's contribution to evolutionary synthesis, see John Beatty, "Julian Huxley and the Evolutionary Synthesis," in Waters and Van Helden, eds., *Julian Huxley*, 181–93. On the history of the evolutionary synthesis, see Smocovitis, *Unifying Biology*.

108. Huxley, *Evolution: The Modern Synthesis*, chap. 10 ("Evolutionary Progress").

109. With time, UNESCO became Huxley's virtual laboratory for testing his ideas about cultural evolution. Thus, he interpreted increasing internationalism and scientific planning as they were taking shape in the United Nations and its specialized agencies as evidence of a movement toward the simultaneous complexity and unity of the social-political life. See Sebastián Gil-Riaño, "Historicizing Anti-Racism: UNESCO's Campaigns against Race Prejudice in the 1950s" (PhD diss., University of Toronto, 2014).

110. Julian Huxley, *UNESCO's Philosophy*, 11, 55.

111. Huxley, *UNESCO's Philosophy*, 8, 5, 72, 73.

112. Huxley, *UNESCO's Philosophy*, 42.

113. "Scientific and Cultural History," Resolution 5.7, *Records of the General Conference of the United Nations Educational and Cultural Organization, Second Session, Mexico, 1947* (Paris: UNESCO, 1947), 27.

114. "Verbatim Report of Talk by Dr. Huxley at the Sorbonne University, Paris, on Thursday, 26 February, 1948, at 9.0. P.M.," 21, Julian Huxley Papers, box 66, folder 7.

115. UNESCO, *A Handbook for the Improvement of Textbooks and Teaching Materials as Aids in International Understanding* (Paris: UNESCO, 1949), 2. On UNESCO's revision of history textbooks, see Ines Dussel and Christian Ydesen, "UNESCO and the Improvement of History Textbooks in Mexico, 1945–1960," in *A History of UNESCO: Global Actions and Impacts*, ed. Poul Duedahl (London: Palgrave Macmillan, 2016), 231–56.

116. Julian Huxley to Director General, memorandum, Paris, January 17, 1949, Julian Huxley Papers, box 118, folder 3.

CHAPTER FIVE

1. Alvin M. Weinberg, "Impact of Large-Scale Science on the United States," *Science* 134, no. 3473 (1961): 161–64.

2. See, e.g., James H. Capshew and Karen Rader, "Big Science: Price to the Present," *Osiris* 7 (1992): 3–25; Peter Galison and Bruce Hevly, eds., *Big Science: The Growth of Large-Scale Research* (Stanford, CA: Stanford University Press, 1992); Catherine Westfall, "Rethinking Big Science: Modest, Mezzo, Grand Science and the Development of the Bevalac, 1971–1993," *Isis* 94 (2003): 30–56; and Daniel J. Kevles, "Big Science and Big Politics in the United States: Reflections on the Death of the SSC and the Life of the Human Genome Project," *Historical Studies in the Physical and Biological Sciences* 27, no. 2 (1997): 269–97.

3. As the preamble to the UNESCO constitution states: "Since wars begin in the *minds* of men, it is in the *minds* of men that the defences of peace must be constructed." See http://portal.unesco.org/en/ev.php-URL_ID=15244&URL_DO=DO_TOPIC&URL_SECTION=201.html.

4. Throughout the 1950s and 1960s the United Nations actively supported studies of intergroup communication as a means of realizing peace, enlisting the aid of social scientists to understand small-group interaction. See Jamie

Cohen-Cole, *The Open Mind: Cold War Politics and the Sciences of Human Nature* (Chicago: University of Chicago Press, 2014).

5. See Lorraine Daston, "The Immortal Archive: Nineteenth-Century Science Imagines the Future," in *Science in the Archives: Pasts, Presents, Futures*, ed. Lorraine Daston (Chicago: University of Chicago Press, 2017), 159–82.

6. Quoted in Daston, "The Immortal Archive," 160.

7. On botany as another example of nineteenth-century Big Science, see Londa Schiebinger and Claudia Swan, eds., *Colonial Botany: Science, Commerce, and Politics in the Early Modern World* (Philadelphia: University of Pennsylvania Press, 2004).

8. On Huxley's collaboration with Wells, see chapter 4 above.

9. See Ross, *H. G. Wells's World Reborn*.

10. Julian Huxley, "Notes on the Scientific and Cultural History of Mankind," May 1948, Julian Huxley Papers, box 118, folder 3.

11. Alvin M. Weinberg, "Criteria for Scientific Choice," *Minerva* 1 (1963): 159–71. For the larger political context of Weinberg's discussion, see Aronova, "The Congress for Cultural Freedom."

12. Weinberg, "Criteria for Scientific Choice," 159. On the PSAC, see chapter 6 below.

13. Weinberg, "Impact of Large-Scale Science on the United States," 161.

14. Alvin M. Weinberg, *Reflections on Big Science* (New York: Pergamon, 1967), 40.

15. For a detailed account of the planning and implementation of the History of Mankind project, see Poul Duedahl, "Selling Mankind: UNESCO and the Invention of Global History, 1945–1976," *Journal of World History* 22, no. 1 (2011): 101–33.

16. See, e.g., "Verbatim Report of Talk by Dr. Huxley."

17. Notes by Julian Huxley, Julian Huxley Papers, box 118, folder 3.

18. Joseph Needham to Julian Huxley, October 30, 1948, Needham Papers, ser. D, folder 161.

19. Patrick Petitjean and Heloisa Maria Bertol Domingues, "1947–1950: Quand l'Unesco a cherché à se démarquer des histoires européocentristes: Le projet d'une histoire scientifique et culturelle de l'humanité" (2007), https://halshs.archives-ouvertes.fr/halshs-00166355.

20. Joseph Needham, *Chinese Science* (London: Pilot, 1945).

21. Joseph Needham to Julian Huxley, October 13, 1948, Needham Papers, ser. D, folder 161. He added: "Ascribing a leading role to the material factors—geographic, climatic, social and economic—I feel to be quite indispensable. How interesting it will be to see what happens." This was the beginning of what became the monumental series *Science and Civilisation in China* (Cambridge: Cambridge University Press, 1954–). After Needham's death in 1995, the project was being managed by the Needham Research Institute. For the list of volumes to date, see http://www.nri.cam.ac.uk/science.html.

22. Needham Papers, ser. D, folder 161.

23. Needham Papers, ser. D, folder 161.

24. Petitjean and Domingues, "1947–1950."

25. *Actes de la 3ᵉ session de la conference générale* (Paris: UNESCO, 1948), 3:358.

26. Lucien Febvre, "Report of the General Conference in Beirut to the French National Commission," February 26, 1949, 9–13, Needham Papers, ser. D, folder 161.

27. Lucien Febvre, "Face au vent: Manifeste des Annales nouvelles," *Annales: Économies, sociétés, civilisations* 1, no. 1 (1946): 1–8, and " 'L'histoire, c'est la paix?" *Annales: Économies, sociétés, civilisations* 11, no. 1 (1956): 51–3. See also Lucien Febvre and François Crouzet, *Nous sommes des sang-mêlés: Manuel d'histoire de la civilisation française* (Paris: A. Michel, 2012).

28. Febvre, "Report of the General Conference in Beirut to the French National Commission."

29. "Rapport de M. Lucien Febvre," 1949, Julian Huxley Papers, box 118, folder 3.

30. "Rapport de M. Lucien Febvre."

31. "Report of the Committee of Experts Responsible for Preparing the Plan of the Scientific and Cultural History of Mankind," Unesco House, Paris, December 12–16, 1949, JSHP, box 118, folder 3, and "Scientific and Cultural History of Mankind: Individual Comments on the Plan Prepared by the Committee of Experts," May 25, 1950, http://unesdoc.unesco.org/images/0015/001551/155103mb.pdf.

32. The move was reflected in the new practice of appointing representatives of governments rather than exemplars of scholarship and personal integrity as members of the UNESCO Executive Board and UNESCO national commissions. On the "one worldism" of UNESCO in its early years, see Glenda Sluga, "UNESCO and the (One) World of Julian Huxley," *Journal of World History* 21, no. 3 (2010): 393–418. On the imperial mind-set in the early history of the United Nations and its agencies, see Mark Mazower, *The End of Empire and the Ideological Origins of the United Nations* (Princeton, NJ: Princeton University Press, 2009).

33. Quoted in Petitjean and Domingues, "1947–1950."

34. "Evaluation of Communist Infiltration of UNESCO: Report, Central Intelligence Agency," February 7, 1947, quoted in Duedahl, "Selling Mankind," 107.

35. The project's outcome was published as Ernest Barker, George Clark, and P. Vaucher, eds., *The European Inheritance*, 3 vols. (Oxford: Clarendon, 1954).

36. The commission included R. E. Mortimer Wheeler, Harold Idris Bell, R. Syme, C. F. C. Hawkes, and Charles Webster. For further discussion and references, see Elena A. Aronova, "Studies of Science before 'Science Studies': Cold War and the Politics of Science in the U.S., U.K., and U.S.S.R., 1950s–1970s" (PhD diss., University of California, San Diego, 2012).

37. https://atom.archives.unesco.org/international-commission-for-history-of-scientific-and-cultural-development-of-mankind.

38. Julian Huxley Papers, box 118, folder 3.

39. Petitjean and Domingues, "1947–1950."

40. Cited in Duedahl, "Selling Mankind," 117. On the critique of new postwar international bureaucracies as reproducing the imperial order, see Michael

Adas, *Machines as the Measure of Men: Science, Technology and the Ideology of Western Dominance* (Ithaca, NY: Cornell University Press, 1989); Joseph Morgan Hodge, *Triumph of the Expert: Agrarian Doctrines of Development and the Legacies of British Colonialism* (Athens: Ohio University Press, 2007); and Richard Jolly, Louis Emmerij, Dharam Ghai, and Frederic Lapeyre, *UN Contributions to Development Thinking and Practice* (Bloomington: Indiana University Press, 2004).

41. Chloé Maurel, "L'histoire de l'humanité de l'UNESCO (1945–2000)," *Revue d'histoire des sciences humaines* 22, no. 1 (2010): 161–98, 177.

42. Kelly Ann Mulroney, "Team Research and Interdisciplinarity in French Social Science, 1925–1952" (PhD diss., University of Virginia, 2000).

43. André-Georges Haudricourt, "De l'origine de l'attelage moderne," *Annales d'histoire économique et sociale* 8 (1936): 515–22.

44. Lucien Febvre, "Amérique du Sud: Un champ privilégié d'études," *Annales d'histoire économique et sociale* 2 (1929): 258–78.

45. In 1948, the *Annales* devoted an entire double issue to Latin America. See "A travers les Amérique Latine," special issue, *Annales: Economies, sociétés, civilisations*, vol. 3, no. 4 (1948).

46. For Braudel, Brazil became the site from which he arrived at his historiographic concept of the Mediterranean as part of a larger Atlantic world. As Charles Morazé, the coeditor of the *Annales* in the years immediately following World War II who also spent the years 1949–51 living in Brazil, remarked in 1985: "The Mediterranean world . . . the idea that has enjoyed recent success was already known fifty years ago [in] Algeria, the Mediterranean, Brazil: countries near or below the tropics." Quoted in Pierre Daix, *Braudel* (Paris: Flammarion, 1995), 115. Among the *Annales* historians, this perspective not only stimulated an intense interest in Latin America and its history but also prompted collaborations between Brazilian and other Latin American intellectuals. See Guy Martinère, *Aspects de la coopération franco-brésilienne: Transplantation culturelle et stratégie de la modernité* (Grenoble: Presses universitaires de Grenoble, 1982).

47. "Lucien Febvre et la Commission international pour une histoire du development scientifique et culturel de l'humanité," UNESCO Archives, AG 8 Secretariat Records, SCHM1—International Commission, ser. 6, Journal of World History, box SCHM 52.

48. Karl Jettmar, "Les plus anciennes civilizations d'éleveurs des steppes d'Asie Centrale," *Cahiers d'histoire mondiale* 1 (1954): 760–83.

49. Lucien Febvre, "The Editor's Role," *Cahiers d'histoire mondiale* 1, no. 1 (1953): 198–201, 200.

50. For an insider account of Febvre's work on the journal, see Charles Morazé, *Un historien engagé: Mémoires* (Paris: Fayard, 2007). Febvre died the following year, aged seventy-eight.

51. "Lucien Febvre et la Commission international pour une histoire du development scientifique et culturel de l'humanité."

52. "Rapport de M. Febvre, Mai 1949," *Cahiers d'histoire mondiale* 1, no. 4 (1953): 954–61.

53. Febvre had adopted many of these practices from his mentor, Henri Berr, and his projects. See chapter 1 above.

54. *Cahiers d'histoire mondiale* was published from 1953 to 1972.

55. M. G. S. Hodgson, "The Hemispheric Interregional History as an Approach to World History," *Cahiers d'histoire mondiale* 1, no. 3 (1954): 715–23.

56. On Breasted, see chapter 3 above. On Hodgson, see Edmund Burke and Robert J. Mankin, eds., *Islam and World History: The Ventures of Marshall Hodgson* (Chicago: University of Chicago Press, 2018).

57. The following discussion of Hodgson is drawn from Michael Geyer, "The Invention of World History from the Spirit of Nonviolent Resistance," in Burke and Mankin, eds., *Islam and World History*, 55–81.

58. Geyer, "The Invention of World History," 61.

59. Edmund Burke III, "Marshall G. S. Hodgson and the Hemispheric Interregional Approach to World History," *Journal of World History* 6, no. 2 (1995): 237–50, 246.

60. "Plan of a History of the Scientific and Cultural Development of Mankind," *Cahiers d'histoire mondiale* 1 (1953): 230.

61. See the discussion of the process in Katja Naumann, "Avenues and Confines of Globalizing the Past: UNESCO's International Commission for a 'Scientific and Cultural History of Mankind' (1952–1969)," in *Networking the International System: Global Histories of International Organizations*, ed. Madeleine Herren (Heidelberg: Springer, 2014), 187–200, 195.

62. See Louis Gotschalk to Marshall Hodgson, September 9, 1953, quoted in Geyer, "The Invention of World History," 65.

63. Hodgson, "The Hemispheric Interregional History as an Approach to World History," 248.

64. Geyer, "The Invention of World History," 65.

65. Hodgson published another essay in the *Cahiers*. See M. G. S. Hodgson, "The Unity of Later Islamic History," *Cahiers d'histoire mondiale* 5 (1960): 879–914.

66. Edward Said, *Orientalism* (New York: Pantheon, 1978).

67. On the history of Soviet participation in UNESCO, see Aigul Kulnazarova, "Debating International Understanding in the Eastern World: UNESCO and the Soviet Union," in *UNESCO without Borders: Educational Campaigns for International Understanding*, ed. Aigul Kulnazarova and Christian Ydesen (London: Routledge, 2017), 256–74.

68. A. A. Zvorykin to Guy Métraux, October 5, 1955, RGANI, fond 5, opis' 30, delo 133.

69. See Louis H. Porter, "Cold War Internationalisms: The USSR in UNESCO, 1945–1967" (PhD diss., University of North Carolina at Chapel Hill, 2018).

70. Quoted from a memo dated September 5, 1955, by the head of the Central Committee's Division of Science and Culture, Alexei Roumiantsev, in which "the desirability for Soviet philosophers and social scientists to join international associations" was stressed. RGANI, fond 5, opis' 30, delo 524.

71. On the post–World War II expansion of the history profession in the Soviet Union, see Roger D. Markwick, *Rewriting History in Soviet Russia: The Politics of Revisionist Historiography, 1956–1974* (New York: Palgrave, 2001).

72. See, e.g., Kassof, "A Book of Socialism."

73. Quoted in William Benton, "The Great Soviet Encyclopedia," *Yale Review*, n.s., 47 (1958): 552–68, 553. On Sergei Vavilov, see Alexei Kojevnikov, "President of Stalin's Academy: The Mask and Responsibility of Sergei Vavilov," *Isis* 87, no. 1 (1996): 18–50.

74. For biographical detail, see K. A. Shchadilova and T. Z. Kozlova, "Riadom s nim liudi stanovilis' luchshe (k stoletiiu so dnia rozhdeniia A. A. Zvorykina)," *Sotsiologicheskie issledovaniia* 12 (2001): 81–83; and ARAN, fond 425, opis' 3, delo 19 (Zvorykin's personnel file).

75. Shchadilova and Kozlova, "Riadom s nim liudi stanovilis' luchshe."

76. On the history of this short-lived institute, see A. N. Dmitriev, "Institut istorii nauki i tekhniki v 1932–1936 gg.," *Voprosy istorii estestvoznaniia i tekhniki* 1 (2002): 3–41; and Yu. I. Krivonosov, "Institut istorii nauki i tekhniki: Tridtsatye-gromovye, rokovye . . . ," *Voprosy istorii estestvoznaniia i tekhniki* 1 (2002): 42–75.

77. Yu. I. Krivonosov, "Vopros ob organizatsii institute sniat s obsuzhdeniia," *Voprosy istorii estestvoznaniia i iekhniki* 2 (2006): 114–27. See also S. S. Ilizarov, "IIET v 1953 godu," *Voprosy istorii estestvoznaniia i tekhniki* 4 (1993): 88–116.

78. N. I. Bukharin to Gleb Krzhizhanovsky and Nikolai Gorbunov, June 1, 1936, reprinted in Krivonosov, "Institut istorii nauki," 68.

79. A. A. Zvorykin, "Iskorenit' do kontsa posledstviia trotskistsko-bukharinskogo vreditel'stva na fronte istorii nauki i tekhniki," *Vestnik AN SSSR* 4–5 (1937): 15–24.

80. Shchadilova and Kozlova, "Riadom s nim liudi stanovilis' luchshe." Zvorykin's personnel file (ARAN, fond 425, opis' 3, delo 19) does not mention his expulsion from the Party during the purges of 1938.

81. Zvorykin also suggested using the History of Mankind and other UNESCO projects to collect information on the political affiliations and attitudes of American and European scientists and scholars to be organized into a centralized index card database (*kartoteka*).

82. A. A. Zvorykin to P. N. Pospelov, Dec 24, 1955, RGANI, fond 5, opis' 30, delo 524, list 48. Pospelov's handwritten note on the letter reads "Agreed." RGANI, fond 5, opis' 30, delo 524, list 48.

83. Duedahl, "Selling Mankind." Crucial to the decision was the support of the French members of the commission, especially Charles Morazé, with whom Zvorykin had developed a personal friendship. See Morazé, *Un historien engagé*.

84. Report by A. A. Zvorykin, January 1960, RGANI, fond 5, opis' 33, delo 144.

85. A. A. Zvorykin to D. T. Shepilov, May 29, 1956, RGANI, fond 5, opis' 35, delo 38.

86. For a discussion of the role of *Vestnik* in Khrushchev's de-Stalinization campaign, see Porter, "Cold War Internationalisms."

87. A. A. Zvorykin, "Otchet o poezdke v Parizh na soveshchanie komissii UNESKO, 5–25 maia 1956," n.d., RGANI, fond 5, opis' 35, delo 38.

88. A. A. Zvorykin, Memo to the Central Committee, October 12, 1956, RGANI, fond 5, opis' 35, delo 38.

89. A. A. Zvorykin, "Ot redaktsii," *Vestnik istorii mirovoi kul'tury* 1 (1957): 3–8, 4.

90. Marshal Khodzhson, "Mezhregional'naia istoriia polusharii kak metod izucheniia mirovoi istorii," *Vestnik istorii mirovoi kul'tury* 1 (1957): 9–16.

91. Zvorykin, "Ot redaktsii," 4.

92. E. M. Zhukov et al., eds., *Mirovaia istoriia v desiati tomakh*, 10 vols. (Moscow: Gosudarstvennoe izdatel'stvo politicheskoi literatury, 1955), vol. 1.

93. E. M. Zhukov, "O printsypakh napisaniia 'Vsemirnoi istorii,'" *Vestnik istorii mirovoi kul'tury* 1 (1957): 17–23, 20.

94. Robert Browning, review of *Vestnik istorii mirovoi kul'tury*, *Anglo-Soviet Journal* 28, no. 2 (1958): 26–28.

95. On Gandhi's suggestion, in May 1955 the Soviet Committee for Solidarity with the Countries of Asia was established to promote the ideals and ideas of Soviet-Asian unity. For Khrushchev's vision of Asian and Soviet cultural politics with regard to decolonized Asian states, see Masha Kirasirova, "'Sons of Muslims' in Moscow: Soviet Central Asian Mediators to the Foreign East, 1955–1962," *Ab Imperio* 4 (2011): 106–32.

96. The demand for such studies could be discerned from the speeches of Party ideologues. For instance, addressing a Party congress in February 1956, the Politburo member Anastas Mikoyan stated: "Since our contacts with the East are growing and becoming stronger, our interest . . . in the political and cultural ties with the countries of the East has grown immeasurably." Anastas Mikoyan, "Speech by Anastas Mikoyan at the 20th Party Congress of the Communist Party of the Soviet Union, Feb[ruary] 16, 1956," quoted in Elisabeth Leake and Leslie James, eds., *Decolonisation and the Cold War: Negotiating Independence* (London: Bloomsbury, 2015), 154.

97. Tolz, *Russia's Own Orient*, 21.

98. B. A. Vvedenskii, Memo to the Central Committee, June 24, 1958, RGANI, fond 5, opis' 35, delo 77. Zvorykin resigned from his position as the deputy editor of the *BSE* in 1956 to concentrate on his work on the History of Mankind project.

99. Louis Gottschalk, "Writing World History," *History Teacher* 2, no. 1 (1968): 17–23.

100. Quoted in Katja Naumann, "Decentering World History: Marshall Hodgson and the UNESCO Project," in Burke and Mankin, eds., *Islam and World History*, 82–101, 88.

101. Naumann, "Avenues and Confines of Globalizing the Past."

102. R. C. Majumdar to Ralph Turner, July 30, 1958, UNESCO Archives, AG 8 Secretariat Records, SCHM1, ser. 2 Implementation of Scheme, box SCHM 19.

103. Among the critics were the Syrian historian Constantine K. Zyrayk and the Mexican scholar Silvio Zavala. For details, see Naumann, "Avenues and Confines of Globalizing the Past," and "Decentering World History."

104. R. C. Majumdar to Ralph Turner, July 9, 1958, UNESCO Archives, AG 8 Secretariat Records, SCHM1, ser. 2 Implementation of Scheme, box SCHM 19.

105. Ralph Turner to Paulo E. de Berredo Carneiro, July 17, 1958, UNESCO Archives, AG 8 Secretariat Records, SCHM1, ser. 2 Implementation of Scheme, box SCHM 19.

106. Ralph Turner to A. Zvorykin, August 2, 1958 (in Russian translation), RGANI, fond 5, opis' 35, delo 77.

107. G. A. Zhukov, "Kuda idet IuNESKO?" *Pravda*, November 13, 1958, 4.

108. On changing Soviet attitudes toward UNESCO, see I. V. Gaiduk, "Sovetskii soiuz I IuNESKO v gody 'kholodnoi voiny,' 1945–1967," *Novaia i noveishaia istoriia* 1 (2007): 20–34.

109. Duedahl, "Selling Mankind."

110. John H. Plumb, "A Great Story Left Untold," *New York Times*, August 1, 1965. Plumb also noted: "I don't often wish I were as rich as Paul Getty. Today I do. I want to buy time on every commercial radio and TV from Patagonia to the North Cape, to hire sky-writing planes in all the world's capitals, to take pages of advertising in all the world's press, just to say how awful, how idiotic is the second volume of UNESCO's projected six-volume 'History of Mankind.' "

111. For an insightful account of how Gottschalk involved graduate students and young scholars in the project, see Naumann, "Avenues and Confines of Globalizing the Past."

112. In the talk at the Sorbonne in which he first announced his plan for a scientific and cultural history project, Huxley mused about the advances in information-processing technology at the Massachusetts Institute of Technology, referring, in particular, to analog computers and microfilm-based information processing as ways to manage information efficiently. "Verbatim Report of Talk by Dr. Huxley," 17.

CHAPTER SIX

1. Lance Davis, "The New Economic History II: Professor Fogel and the New Economic History," *Economic History Review* 19, no. 3 (1966): 657–63, 657.

2. For an overview of quantitative approaches to historical research in the nineteenth century, see William O. Aydelotte, *Quantification in History* (Reading, MA: Addison-Wesley, 1971). On the use of statistical methods and tools in history, see Daniel Rosenberg and Anthony Grafton, *Cartographies of Time: A History of the Timeline* (Princeton, NJ: Princeton Architectural Press, 2010). On the rise of statistics and its appeal to historians in the nineteenth century, see Theodore M. Porter, *The Rise of Statistical Thinking, 1820–1900* (Princeton, NJ: Princeton University Press, 1986); and M. Norton Wise, "How Do Sums Count? On the Cultural Origins of Statistical Causality," in *The Probabilistic Revolution, 1800–1930: Dynamics of Scientific Development*, vol. 1, *Ideas in History*, ed. Lorraine Daston, Michael Heidelberger, and Lorenz Krüger (Cambridge, MA: MIT Press, 1986), 395–425, and "On the Relation of Physical Science to History in Late Nineteenth Century Germany," in *Functions and Uses of Disciplinary Histories*, ed. Loren Graham, Wolf Lepenies, and Peter Weingart (Dordrecht: Reidel, 1983), 3–34.

3. Bernal distanced himself from the Communist Party in the late 1920s and did not publicize his subsequent relationship with it; thus, the circumstances of his relationship with the Party in the 1920s and 1930s are not clear. He claimed to have lost his card in 1933, but he maintained close links to the Party leadership throughout his life. See Brenda Swann and Francis Aprahamian, eds., *J. D. Bernal: A Life in Science and Politics* (London: Verso, 1999), 60–64. Other major sources on Bernal's life and work are Andrew Brown, *J. D. Bernal: The Sage*

of Science (Oxford: Oxford University Press, 2005); Maurice Goldsmith, *Sage: A Life of J. D. Bernal* (London: Hutchinson, 1980); and Werskey, *The Visible College*.

4. J. D. Bernal, "Science and Society," *The Spectator*, July 11, 1931, 43–44.

5. On Crowther and the tours to the Soviet Union, see chapter 4 above.

6. Dorothy Mary Crowfoot Hodgkin, "John Desmond Bernal, 10 May 1901–15 September 1971," *Biographical Memoirs of Fellows of the Royal Society*, vol. 26 (1980), https://doi.org/10.1098/rsbm.1980.0002.

7. Quoted in Hodgkin, "John Desmond Bernal," 67.

8. J. D. Bernal, *The Social Function of Science* (1939), 4th ed. (Cambridge, MA: MIT Press, 1967). For an overview of Bernal's career as an information scientist, see Dave Muddiman, "Red Information Scientist: The Information Career of J. D. Bernal," *Journal of Documentation* 59, no. 4 (2003): 387–409.

9. Bernal, *The Social Function of Science*, 292–308. As Alex Csiszar has demonstrated, the scientific paper was itself a relatively recent phenomenon. See Alex Csiszar, *The Scientific Journal: Authorship and the Politics of Knowledge in the Nineteenth Century* (Chicago: University of Chicago Press, 2018).

10. See Bernal, *The Social Function of Science*, 449–57 (app. 8, "A Project for Scientific Publication and Bibliography"), 449.

11. The US Science Service was established in the 1920s under the auspices of the National Research Council as a solution to the fragmentation of science as well as a means of bolstering an image of American science in Europe. See Marcel Chotkowski LaFollette, *Science on the Air: Popularizers and Personalities on Radio and Early Television* (Chicago: University of Chicago Press, 2008); and L. C. Mendes, "Enhancing Lives through Information and Technology: Watson Davis's Project for Information Organisation and Dissemination," *Proceedings of the Association for Information Science and Technology* 53 (2016): 1–7.

12. Bernal, *The Social Function of Science* (London: George Routledge and Sons, 1939), 312.

13. See Alistair Black, Helen Plant, and Dave Muddiman, eds., *The Early Information Society: Information Management in Britain Before the Computer* (Burlington, VT: Ashgate, 2007).

14. See Alex Wright, *Cataloging the World: Paul Otlet and the Birth of the Information Age* (Oxford: Oxford University Press, 2014).

15. H. G. Wells, *World Brain* (London: Methuen & Co., 1938). On H. G. Wells see chapter 4 above.

16. See Dave Muddiman, "Science, Industry, and the State: Scientific and Technical Information in Early-Twentieth-Century Britain," in Black, Plant, and Muddiman, eds., *The Early Information Society*, 55–78, 67.

17. The British Council's Science Committee was founded in early 1941 to promote British science abroad. Crowther was employed at the British Council from 1941 to 1946. See Allan Jones, "J. G. Crowther's War: Institutional Strife at the BBC and British Council," *British Journal for the History of Science* 49, no. 2 (2016): 259–78. See also Muddiman, "Red Information Scientist."

18. J. G. Crowther, "Scientific Information," in *Report of the Proceedings of the Eighteenth Conference of the Association of Special Libraries and Information Bureau* (London: ASLIB, 1943), 14–20, 15, 20.

19. "Publication and Classification of Scientific Knowledge," *Nature* 160 (November 8, 1947): 649–50. See also J. D. Bernal, "Information Science as an Essential in the Progress of Science" (1946), in *The Freedom of Necessity* (London: Routledge & Kegan Paul, 1949), 226–53, and "The Supply of Information to the Scientist: Some Problems of the Present Day," *Journal of Documentation* 13, no. 4 (1957): 195–208.

20. Muddiman, "Red Information Scientist."

21. Muddiman, "Science, Industry, and the State," 68.

22. J. D. Bernal, "Provisional Scheme for Central Distribution of Scientific Publications," in *The Royal Society Scientific Information Conference, 21 June–2 July 1948, London: Reports and Papers Submitted* (London: Royal Society, 1948).

23. Quoted in Muddiman, "Science, Industry, and the State," 70.

24. On Polanyi, see Mary Jo Nye, *Michael Polanyi and His Generation: Origins of the Social Construction of Science* (Chicago: University of Chicago Press, 2011).

25. On the SFS, see Wolfe, *Freedom's Laboratory.*

26. John R. Baker and A. G. Tansley, "The Course of the Controversy on Freedom in Science," *Nature* (October 26, 1946): 574–76.

27. See Wolfe, *Freedom's Laboratory*; and Aronova, "The Congress for Cultural Freedom." See also chapter 4 above.

28. On the membership of the SFS, see Nye, *Michael Polanyi and His Generation*, 205.

29. Baker and Tansley, "The Course of the Controversy on Freedom in Science," 576.

30. G. P. Thomson and John R. Baker, "Proposed Central Publication of Scientific Papers," *Nature* 161 (1948): 771–72.

31. See the detailed discussion of the media campaign against Bernal in Harry East, "Professor Bernal's 'Insidious and Cavalier Proposals': The Royal Society Scientific Information Conference, 1948," *Journal of Documentation* 54, no. 3 (1998): 294–302. For a discussion of the Cold War context of these attacks, see Wolfe, *Freedom's Laboratory.*

32. J. D. Bernal, "The Biological Controversy in the Soviet Union and Its Implications," *Modern Quarterly* 4, no. 3 (1949): 203–17. On Lysenko and the crisis in Soviet genetics, see chapter 4 above. For Polanyi's role in the anti-Communist Congress for Cultural Freedom, , see Wolfe, *Freedom's Laboratory.*

33. On Bernal's role in World Federation of Scientific Workers, see Patrick Petitjean, "The Joint Establishment of the World Federation of Scientific Workers and of UNESCO After World War II," *Minerva* 46, no. 2 (2008): 247–70.

34. See Brown, *J. D. Bernal*, esp. chap. 19 ("Marxist Envoy").

35. I. D. Rozhanskii, "Dzh. D. Bernal (K 50-letiiu so dnia rozhdeniia)," *Uspekhi fizicheskikh nauk* 45, no. 2 (1951): 169–94. I. D. Rozhanskii was the high-level administrator at the Academy of Sciences foreign office who made arrangements for Bernal during his visits to the Soviet Union. See ARAN, fond 579, opis' 3-L, delo 20 (Bernal's personal file as a foreign member of the Soviet Academy of Sciences).

36. According to a report by the head of staff of the foreign office of the Academy of Sciences, Sergei Vavilov was the first scientist Bernal indicated he

wanted to meet during his visit in 1949. See I. Rozhanskii, "Nekotorye zamechaniia ot vstrech s prof. Bernalom," n.d., ARAN, fond 579, opis' 3-L, delo 20, listy 10–13.

37. For a review of early Soviet proposals to centralize information, see, e.g., M. J. Goldman, "Reform in the System of Scientific Publication," *Science* 80 (October 26, 1934): 380–81.

38. See G. S. Tsypkin, ed., *Aleksandr Nikolaevich Nesmeyanov—organizator nauki* (Moscow: Nauka, 1996), 181.

39. See, e.g., György Rózsa, *Scientific Information and Society* (The Hague: Mouton, 1973).

40. The Soviet atomic bomb project set an important precedent. In 1942, the Soviet physicist Georgii Flerov rightly deduced the Allies' atomic bomb project by the sudden disappearance of publications on nuclear fission in physics journals and wrote directly to Josef Stalin to warn him of the danger. In 1945, a sanitized history of the Manhattan Project authored by a Princeton physicist, Henry DeWolf Smyth (the "Smyth Report"), provided a great template for the Soviet scientists involved in the atomic bomb project. Even though the report lacked technical details, it gave valuable clues about what kinds of plants to build, what methods worked, how much work was involved, and so forth. The Smyth Report was immediately translated into Russian after its release in the United States in 1945. See Michael D. Gordin, *Red Cloud at Dawn: Truman, Stalin, and the End of the Atomic Monopoly* (New York: Farrar, Straus & Giroux, 2010).

41. On the history of the Institute for Scientific Information, later renamed the All-Union Institute for Scientific and Technical Information (see below), see A. I. Chernyi, *Vserossiiskii institut nauchnoi i tekhnicheskoi informatsii: 50 let sluzheniia nauke* (Moscow, 2005). See also Michael D. Gordin, *Scientific Babel: How Science Was Done Before and After Global English* (Chicago: University of Chicago Press, 2015), 250–51.

42. Quoted in Muddiman, "Red Information Scientist," 399.

43. Eugene Garfield and Robert Hayne, "Needed: A National Science Intelligence and Documentation Center," December 1955, Correspondence and Reports: 1955–57, IGY Papers. See also R. C. Peavey to Eugene Garfield, August 4, 1956, Correspondence and Reports: 1955–57, IGY Papers. For a discussion of this episode, see Elena Aronova, "Geophysical Datascapes of the Cold War: Politics and Practices of the World Data Centers in the 1950s and 1960s," *Osiris* 32 (2017): 307–27.

44. Colin Burke argues that Garfield, who was born in 1925 into a working-class family of second-generation Jewish immigrants and had some relatives involved in radical organizations and labor unions in the 1930s, was committed to ideals of social justice, which explains, e.g., his sympathy toward Bernal. See Colin B. Burke, *America's Information Wars: The Untold Story of Information Systems in America's Conflicts and Politics from World War II to the Internet Age* (Lanham, MD: Rowman & Littlefield, 2018), 287–89.

45. Eugene Garfield, "Citation Indexes—New Paths to Scientific Knowledge," *Chemical Bulletin* 43, no. 4 (1956): 11–12.

46. Garfield proposed establishing a newspaper for science, the *Daily Scientist*, that would follow the format of the *New York Times*. In addition to original research articles, each issue would include reviews, a daily updated author

bibliography, a subject index, and a citation index. As its name implied, the scientific newspaper was not intended for keeping and collecting, like scientific journals: "The philosophy behind a daily dissemination technique is that the information comes in small segments. The daily newspaper is quickly scanned and then discarded." Eugene Garfield, "Application of Newspaper Publishing Methods to the Dissemination of Scientific Information" (1962), proposal to the National Science Foundation by the Institute for Scientific Information, Philadelphia, Garfield Papers.

47. Brown, *The Sage of Science*, 364. See also chapter 1 above.

48. E. Garfield, "J. D. Bernal—the Sage of Cambridge," *Current Contents* 19 (1982): 5–14.

49. "ISI History," n.d., Garfield Papers.

50. On the PSAC, see Zuoyue Wang, *In Sputnik's Shadow: The President's Science Advisory Committee and Cold War America* (New Brunswick, NJ: Rutgers University Press, 2008); and chapter 5 above. For a general discussion of the impact of Sputnik on science in the United States, see Robert A. Divine, *The Sputnik Challenge: Eisenhower's Response to the Soviet Satellite* (New York: Oxford University Press: 1993). See also David Kaiser, "The Physics of Spin: Sputnik Politics and American Physicists in the 1950s," *Social Research* 73, no. 4 (2006): 225–52.

51. President's Science Advisory Committee, "The Responsibilities of the Technical Community and the Government in the Transfer of Information: A Report" (Washington, DC: White House, January 10, 1963), 7, Garfield Papers.

52. President's Science Advisory Committee, "The Responsibilities of the Technical Community and the Government," 35.

53. Eugene Garfield, "Citation Indexes for Science," *Science* 122, no. 3159 (1955): 108–11. For the history of the creation of the SCI, see Paul Wouters, "The Creation of the Science Citation Index," in *Proceedings of the 1998 Conference on the History and Heritage of Science Information Systems*, ed. M. E. Bowden, T. Bellardo Hahn, and R. V. Williams (Pittsburgh, PA: Information Today, 1999), 127–36. For a detailed discussion of the origin of the SCI, see Paul Wouters, "The Citation Culture" (PhD diss., University of Amsterdam, 1999).

54. Garfield, "Citation Indexes for Science," 108, 109.

55. On the collaboration between Lederberg and Bernal, see Wouters, "The Citation Culture."

56. J. D. Bernal, "Science Citation Index," *Science Progress* 53, no. 211 (1965): 455–59.

57. Eugene Garfield to Joshua Lederberg, November 10, 1961, Garfield Papers. Garfield thought that machine translation and citation indexing were "still somewhat primitive" tools to cope with the avalanche of publications and that "better methods of mechanizing" were required to collect and process vast amounts of data. He was convinced, however, that "regardless of the imperfections of all of these . . . we can expect that high-speed computers will be able to process . . . material at a cost that is not higher than the input cost of $10,000,000." Ibid.

58. Eugene Garfield, Irving H. Sher, and Richard J. Torpie, *The Use of Citation Data in Writing the History of Science* (Philadelphia: Institute for Scientific Information, 1964).

59. Garfield, Sher, and Torpie, *The Use of Citation Data*, iii.

60. Garfield, "Citation Indexes for Science," 109.

61. See Grafton, *The Footnote*. Steven Shapin and Simon Schaffer argued that one of the characteristic features of the "experimental form of life" that emerged in seventeenth-century England and became a standard norm of science was "the injunction against the ornamental citing of authorities." According to Shapin and Schaffer, the new experimental philosophers at the Royal Society began to use "citations of other writers . . . not as 'judges, but as witnesses,' as 'certificates to attest matters of fact.' " Steven Shapin and Simon Schaffer, *Leviathan and the Air-Pump: Hobbes, Boyle and the Experimental Life* (Princeton, NJ: Princeton University Press, 1985), 68.

62. One possible explanation for Garfield's pitching the SCI as a tool for historians might stem from the fact that early funding for his work came about through national security connections. Presenting the SCI as a tool for historians might have served as a public facade for a project with national security and intelligence connections. Before getting a position at the Johns Hopkins Indexing Project, Garfield had interviewed for a job with the Massachusetts Institute of Technology's chemical information project, which was partly funded by the CIA. See Burke, *America's Information Wars*, 289. Garfield's contacts with the CIA not only proved useful in securing him support from the air force but also continued well into the 1970s. See "Memorandum (Sanitized)," October 1, 1979, General CIA Records (CREST), https://www.cia.gov/library/readingroom/document/cia-rdp90-00509r000100020003-2.

63. See Isaac Asimov, *The Genetic Code* (New York: New American Library, 1962). In *The Use of Citation Data*, Garfield thanked Asimov "for his willingness to submit his book to the liberties of our discussion." Garfield, Sher, and Torpie, *The Use of Citation Data*, iii. Garfield's personal papers contain only later correspondence with Asimov, i.e., from the 1970s and 1980s. His choice of *The Genetic Code* for his experiment might have been motivated by the fact that, in his earlier *Foundation* series, published between 1942 and 1950, Asimov sought to make history scientific and treated it as a science. I thank Patrick McCray for pointing this out to me.

64. Garfield, Sher, and Torpie, *The Use of Citation Data*, 2.

65. For the history of the genetic code, which historicizes the idioms of *code* and *information*, see Lily E. Kay, *Who Wrote the Book of Life? A History of the Genetic Code* (Stanford, CA: Stanford University Press, 2000).

66. Asimov, *The Genetic Code*, xiv.

67. Garfield, Sher, and Torpie, *The Use of Citation Data*, iii, iv.

68. On Franklin's contributions, see Brenda Maddox, *Rosalind Franklin: The Dark Lady of DNA* (New York: HarperCollins, 2002). On the gendered history of genetics, see Evelyn Fox-Keller, *Reflections on Gender and Science* (New Haven, CT: Yale University Press, 1985), and *A Feeling for the Organism: The Life and Work of Barbara McClintock* (San Francisco: W. H. Freeman, 1983).

69. Garfield, Sher, and Torpie, *The Use of Citation Data*, iii. A reviewer noted that the most useful aspect of the citation technique was the possibility of capturing articles overlooked by historians. See Stephen G. Brush, review of *The Use of Citation Data in Writing the History of Science*, *Isis* 56, no. 4 (1965): 487.

70. Garfield, Sher, and Torpie, *The Use of Citation Data*, i–iii.

71. Among other contributions to the International Symposium on Quantitative Measures in the History of Science, Berkeley, CA, August 25–27, 1976, see David Edge's "Quantitative Measures of Communication in Science." See also "Citation Studies of Scientific Specialties," special issue, *Social Studies of Science*, vol. 7, no. 2 (May 1977).

72. Eugene Garfield to Derek de Solla Price, March 6, 1962, Garfield Papers.

73. Derek de Solla Price to Eugene Garfield, March 15, 1962, Garfield Papers.

74. Derek de Solla Price, *Science since Babylon* (New Haven, CT: Yale University Press, 1961).

75. Derek de Solla Price, *Little Science, Big Science* (New York: Columbia University Press, 1963).

76. Price to Garfield, March 15, 1962.

77. Robert K. Merton and Eugene Garfield, foreword to Price, *Little Science, Big Science*, 2nd ed. (New York: Columbia University Press, 1986), ix. For a discussion of the collaboration between Garfield and Price, see Wouters, "The Citation Culture."

78. In the 1970s, Garfield worked on developing computer-produced "historiographs" to visualize Kuhnian revolutions in science and the emergence of new fields. See Eugene Garfield, "Historiographs, Librarianship and the History of Science," *Current Contents* 38 (1974): 134–35, 135. In later years, he developed a method of creating historiographs algorithmically, producing "a genealogic profile of the evolution" of a given topic, idea, or field. He hoped that "algorithmic historiography" would eventually be adopted by historians as a method of producing a quick summary that would facilitate an understanding of "paradigms shifts." Eugene Garfield, A. I. Pudovkin, and V. S. Istomin, "Why Do We Need Algorithmic Historiography?" *Journal of the American Society for Information Science and Technology* 54, no. 5 (2003): 400–412, 400. See also Eugene Garfield, "From the Science of Science to Scientometrics: Visualizing the History of Science with HistCite Software," *Journal of Informetrics* 3 (2009): 173–79. For a discussion, see Loet Leydesdorff, "Eugene Garfield and Algorithmic Historiography: Co-Words, Co-Authors, and Journal Names," *Annals of Library and Information Studies* 57, no. 3 (2010): 248–60.

79. Thomas Kuhn, *Structure of Scientific Revolutions* (1962), 3rd ed. (Chicago: University of Chicago Press, 1996), 178, 192. Kuhn used citation analysis in his project on the history of quantum mechanics. In a letter to Garfield in March 1967, he wrote: "As you know, I am much interested in citation studies of the quantum mechanics literature myself." Thomas S. Kuhn to Eugene Garfield, March 30, 1967, Kuhn Papers, box 2, folder 33. On Kuhn's project "Sources for the History of Quantum Physics," see Anke te Heesen, "Thomas S. Kuhn, Earwitness: Interviewing and the Making of a New History of Science," *Isis* 111, no. 1 (2020): 86–97.

80. Marcel Couturier, *Recherches sur les structures sociales de Châteaudun, 1525–1789* (Paris: SEVPEN, 1969).

81. See James W. Cortada, *The Digital Flood: The Diffusion of Information Technology across the U.S., Europe, and Asia* (New York: Oxford University

Press, 2012). On the complex history of information technologies in France, see Pierre-Éric Mounier-Kuhn, *L'informatique en France de la Seconde Guerre Mondiale au Plan Calcul: L'émergence d'une science* (Paris: Presses de l'Université Paris-Sorbonne, 2010); and Julien Mailland and Kevin Driscoll, *Minitel: Welcome to the Internet* (Cambridge, MA: MIT Press, 2017). For the cultural history of French policies regarding technological modernization after World War II, see Michael Bess, *The Light-Green Society: Ecology and Technological Modernity in France, 1960–2000* (Chicago: University of Chicago Press, 2003).

82. See Michel Fleury and Louis Henry, *Nouveau manuel de dépouillement et d'exploitation de l'état civil ancient* (Paris: Institut National d'Etudes Démographiques, 1965).

83. See the discussion in Charles Tilly, "Quantification in History, as Seen from France," in *The Dimensions of the Past*, ed. Val Lorwin and Jacob Price (New Haven, CT: Yale University Press, 1972), 93–125.

84. Emmanuel Le Roy Ladurie, *Le territoire de l'historien* (Paris, 1973), 1:14.

85. See Fernand Braudel, "History and the Social Sciences: The *Longue Durée*" (1958), *Review: Fernand Braudel Center* 32, no. 2 (2009): 171–203, 202. On the *Annales* school's emphasis on quantitative methods in the 1960s and 1970s, see George Huppert, "The Annales Experiment," in *Companion to Historiography*, ed. Michael Bentley (London: Routledge, 1997); and Burke, *The French Historical Revolution.*

86. J. H. Hexter, "Fernand Braudel and the Monde Braudelien," *Journal of Modern History* 44 (1972): 480–539, 515.

87. See, e. g., Jean Marczewski, "Buts et méthodes de l'histoire quantitative," *Cahiers Vilfredo Pareto* 3 (1964): 127–64. See also Robert Forster, "Achievements of the Annales School," *Journal of Economic History* 38, no. 1 (1978): 58–76.

88. See, e.g., Joy Lisi Rankin, *A People's History of Computing in the United States* (Cambridge, MA: Harvard University Press, 2018). On the early Cold War history of computers in the United States, see Paul N. Edwards, *The Closed World: Computers and the Politics of Discourse in Cold War America* (Cambridge, MA: MIT Press, 1996).

89. Issues of the journal from its founding in 1966 until 2004 are accessible at https://www.jstor.org/journal/comphuma. In 2005, *CHum* was renamed *Language Resources and Evaluation.*

90. See, e.g., Rankin, *A People's History of Computing.*

91. "Computer Programs Designed to Solve Humanistic Problems," *Computers and the Humanities* 1, no. 2 (1966): 39–55.

92. Vern L. Bullough, "The Computer and the Historian: Some Tentative Beginnings," *Computers and the Humanities* 1, no. 3 (1967): 61–64, 64.

93. Charles Tilly, "Computers in Historical Analysis," *Computers and the Humanities* 7, no. 6 (1973): 323–35.

94. On the concern in the 1960s that adapting to computer analysis might reduce or reshape the scope of research, making it more quantitative and excluding nondigital aspects, see Aronova, "Geophysical Datascapes of the Cold War."

95. Isaiah Berlin, "History and Theory: The Concept of Scientific History," *History and Theory* 1, no. 1 (1960): 1–31, 7. On Berlin's role in the

anti-Communist Congress for Cultural Freedom, see Scott-Smith, *The Politics of Apolitical Culture.*

96. See Julian Huxley, "Volume 6 [part] 2: Memorandum II," Julian Huxley Papers, box 119, folder 7.

97. Carl Bridenbaugh, "The Great Mutation," *American Historical Review* 68 (1963): 315–31, 326.

98. During the student protests at the University of California, Berkeley, where punch cards were used for class registration and grading, students chanted: "I am not a number." One student pinned a sign on his chest: "I am a UC student. Please do not bend, fold, spindle or mutilate me" (paraphrasing the warning printed on each IBM card: "Do not fold or bend this card"). See Steven Lubar, " 'Do Not Fold, Spindle or Mutilate': A Cultural History of the Punch Card," *Journal of American Culture* 15 (1992): 43–55; and Bernward Joerges, "Images of Technology in Sociology: Computer as Butterfly and Bat," *Technology and Culture* 31 (1990): 203–27.

99. P. T. P. Oliver, "Citation Indexing for Studying Science," *Nature* 227 (1970): 870, quoted in Alex Csiszar, "Metrics and the Bureaucratic Virtuoso: Robert K. Merton, Eugene Garfield, and Goal Displacement in Science" (paper presented at the workshop " 'The Engine of Modernity': Construing Science as the Driving Force of History," May 2–3, 2017, Columbia University). I thank Alex Csiszar for sharing his paper with me.

100. Eugene Garfield, interview by Robert V. Williams, July 29, 1997, transcript, Garfield Papers.

101. See Slava Gerovitch, *From Newspeak to Cyberspeak: A History of Soviet Cybernetics* (Cambridge, MA: MIT Press, 2002), and " 'Mathematical Machines' of the Cold War: Soviet Computing, American Cybernetics and Ideological Disputes in the Early 1950s," *Social Studies of Science* 31, no. 2 (2001): 253–87. See also Ksenia Tatarchenko, "Cold War Origins of the International Federation for Information Processing," *IEEE Annals of the History of Computing* 32, no. 2 (2010): 46–57.

102. The so-called Lacy-Zaroubin Agreement was signed by the ambassadors of the Soviet Union and the United States in January 1958. See https://librariesandcoldwarculturalexchange.wordpress.com/text-of-lacy-zaroubin-agreement-january-27-1958.

103. For the report of one such delegation, see Melville J. Ruggles and Raynard C. Swank, *Soviet Libraries and Librarianship: Report of the Visit of the Delegation of U.S. Librarians to the Soviet Union, May–June 1961, under the U.S.-Soviet Cultural Exchange Agreement* (Chicago: American Library Association, 1962). For other professional exchanges in this period, see Dora Vargha, "Between East and West: Polio Vaccination across the Iron Curtain in Cold War Hungary," *Bulletin of the History of Medicine* 88 (2014): 319–43; and Allen H. Kassof, "Scholarly Exchanges and the Collapse of Communism," *Soviet and Post-Soviet Review* 22, no. 3 (1995): 263–74. On the Cold War politics of scholarly exchanges, see Yale Richmond, *Cultural Exchange and the Cold War: Raising the Iron Curtain* (University Park: Penn State University Press, 2003); Gabrielle Hecht, ed., *Entangled Geographies: Empire and Technopolitics in the Global Cold War* (Cambridge, MA: MIT Press, 2011); and Wolfe, *Freedom's Laboratory.*

104. "U.S.-U.S.S.R. to Exchange Teams of Special Librarians," *Scientific Information Notes* 6, no. 5 (1964):2. See also Ruggles and Swank, *Soviet Libraries and Librarianship*. On "information pilgrimages" to Moscow as part of President Eisenhower's policy regarding US-Soviet exchanges, see Burke, *America's Information Wars*.

105. Eugene Garfield, interview by Robert V. Williams, July 29, 1997.

106. Gary Brooten, "Soviet Official Admits Red Scientists Face Lag on Information," *Philadelphia Inquirer*, January 28, 1966.

107. Eugene Garfield to N. Arutyunov, February 25, 1966, Garfield Papers.

108. On the ways in which both in the United States and in the Soviet Union the development of electronic computing technology was shaped by the Cold War arms race, see Gerovitch, *From Newspeak to Cyberspeak*; and Edwards, *The Closed World*.

109. The embargo was established by the terms of the Exports Control Act of 1949, adopted by NATO's Coordinating Committee for Multilateral Export Controls. For the historical background, see Michael Mastanduno, "Trade as a Strategic Weapon: American and Alliance Export Control Policy in the Early Postwar Period," *International Organization* 42, no. 1 (1988): 121–50.

110. "IBM Moves into Eastern European Market," *Computerworld*, November 17, 1971, 57.

111. Eugene Garfield to N. Arutyunov, July 14, 1971, Garfield Papers.

112. See Eugene Garfield to G. Vladutz, "Prospects of Exchanging Machine Readable Tapes with USSR (VINITI and GPNTE)," memorandum, August 5, 1977, Garfield Papers.

113. Eugene Garfield to N. Arutyunov, January 19, 1982, Garfield Papers.

114. D. Yu. Panov, "Bystrodeistvuiushchie vychislitel'nye mashiny. (Sostoianie i tendentsii razvitiia)," February 1953 (RGANI, fond 5, opis' 17, delo 512 P-5726, listy 8–27), 18. The US intelligence services produced similar reports. See, e.g., the declassified CIA report "Foreign Computer Capabilities," September 25, 1969, https://www.cia.gov/library/readingroom/docs/DOC_0005577292.pdf.

115. The British report praised VINITI as "the largest scientific information center in the world." See *Scientific and Technical Information in the Soviet Union: Report of the D.S.I.R.-Asib Delegation to Moscow and Leningrad, 7–24 June, 1963* (London: Department of Scientific and Industrial Research, 1964), 39.

116. Eugene Garfield, interview by Robert V. Williams, July 29, 1997.

117. In the 1960s, VINITI employed roughly two thousand employees, not counting outside scientists who worked as part-time abstracters. See Chernyi, *Vserossiiskii institut nauchnoi i tekhnicheskoi informatsii*. See also Aronova, "Geophysical Datascapes of the Cold War."

118. V. V. Nalimov, *Kanatokhodets* (Moscow: Progress, 1992). Biographical details given in this chapter are taken from *Kanatokhodets*, Nalimov's autobiography. On Nalimov's mystical anarchism, see Birgit Menzel, "Vasilij V. Nalimov—ein mystischer Anarchist in der sowjetischen Kybernetik," in *Russland und/als Eurasien: Kulturelle Konfigurationen*, ed. Birgit Menzel and Christine Engel (Berlin: Frank & Timme, 2018), 289–302. On mystical anarchism in

Russia in general, see Bernice Glatzer Rosenthal, "Theatre as Church: The Vision of the Mystical Anarchists," *Russian History* 4, no. 2 (1977): 122–41. On the role of religious mysticism in the Moscow School of Mathematics, see Loren Graham and Jen-Michel Kantor, *Naming Infinity: A True Story of Religious Mysticism and Mathematical Creativity* (Cambridge, MA: Harvard University Press, 2009).

119. On the *sharashka*, see Asif Siddiqi, "Scientists and Specialists in the Gulag: Life and Death in Stalin's *Sharashka*," *Kritika: Explorations in Russian and Eurasian History* 16, no. 3 (2015): 557–88.

120. 'For his first publication from the labor camp, see V.V. Nalimov, "Issledovanie materialov detalei ekskavatorov," *Kolyma* 6–7 (1944): 36–41.

121. G. E. Vladutz, V. V. Nalimov, and N. I. Stiazkhin, "Nauchnaia informatsiia kak odna iz zadach kibernetiki," *Uspekhi fizicheskikh nauk* 69, no. 1 (1959): 13–56.

122. Eugene Garfield, "The ISI-VINITI Connection: Remarks on the 50th Anniversary of VINITI" (paper presented at the Sixth International Conference of Information Society, Intelligent Information Processing and Information Technology, October 16–18, 2002, Moscow), http://garfield.library.upenn.edu/papers/viniti101502.html. Garfield's extensive correspondence with Nalimov from the late 1960s until his death in 1997 can be found in the Garfield Papers.

123. Nalimov had begun to work on a mathematical theory of experiment in the Gulag, using probability theory and mathematical analysis to design experiments in the chemical industry. His first book on the subject earned him a research position at the Institute of Rare Metals in Moscow. See V. V. Nalimov, *Primenenie matematicheskoi statistiki pri analize veshchestva* (Moscow: Fizmatgiz, 1960), translated as *The Application of Mathematical Statistics to Chemical Analysis* by Prasenjit Basu (Oxford: Pergamon, 1963). This first book on experimental design was followed by V. V. Nalimov and N. A. Chernova, *Statisticheskie metody planirovaniia ekstremal'nykh eksperimentov* (Moscow: Nauka, 1965), translated as *Statistical Methods for Design of Extremal Experiments* (n.p.: Foreign Technology Division, Wright-Patterson Air Force Base, 1968), and V. V. Nalimov, *Teoriia eksperimenta* (Moscow: Nauka, 1971), translated as *Theorie des experiments* by A. Kramer et al. (Berlin: VEB Deutscher Landwirtschaftsverlag, 1975), the full exposition of Nalimov's mathematical theory of experiment. On the basis of his work on experimental design, Nalimov was awarded a doctoral degree in technical sciences in 1964. His publication record was so impressive that the Attestation Committee, the state agency overseeing the awarding of advanced academic degrees in the Soviet Union, ignored the fact that he had not completed his undergraduate degree.

124. V. V. Nalimov, Yu. P. Adler, and Yu. V. Granovskii, "Informatsionnaia sistema po matematicheskoi teorii exkperimenta: Pt. 1, Obshchee opisanie sistemy," in *Kiberenetika i dokumantalistika* (Moscow: Nauka, 1966), 138–49.

125. Yu. V. Granovskii, "Naukometriia v Moskovskom universitete," *Upravlenie bol'shimi sistemami* 44 (2013): 67–82; Nalimov, *Kanatokhodets*.

126. Z. B. Barinova et al., "Izuchenie nauchnykh zhurnalov kak kanalov sviazi: Otsenka vklada otdel'nykh stran v mirovoi nauchnyi informatsionnyi potok," *Nauchno-tekhnicheskaia informatsiia*, ser. 2, no. 12 (1967): 3–11. See the

discussion in L. G. Gurjeva and P. S. Wouters, "Scientometrics in the Context of Probabilistic Philosophy," *Scientometrics* 52 (2001): 111–26.

127. On the IIET and Soviet *naukovedenie*, see Elena Aronova, "The Politics and Contexts of Soviet Science Studies (*Naukovedenie*): Soviet Philosophy of Science at the Crossroads," *Studies in East European Thought* 63, no. 3 (2011): 175–202.

128. See Sergei Kara-Murza, "Budem skurpulezno podschityvat' ssylki," blog Tsentra izucheniia krisisnogo obshchestva, April 2, 2015, http://old.centero.ru/opinions/budem-skrupulezno-podschityvat-ssylki-no-pro-sebya-otvergat-eti-lozhnye-sushchnosti.

129. See V. V. Nalimov and Z. M. Mul'chenko, *Naukometriia: Izuchenie razvitiia nauki kak informatsionnogo processa* (Moscow: Nauka, 1969), translated as *Measurement of Science: Study of the Development of Science as an Information Process* by USAF Systems Command Translation Division (n.p.: USAF Systems Command, Foreign Technology Division, 1971).

130. On the rise of techniques of forecasting in the Soviet Union and the Soviet bloc, see Vitezslav Sommer, "Forecasting the Socialist Future: Prognostika in Late Socialist Czechoslovakia," in *The Struggle for the Long-Term in Transnational Science and Politics: Forging the Future*, ed. Jenny Andersson and Eglė Rindzevičiūte (London: Routledge, 2015), 144–68. See also Eglė Rindzevičiūte, *The Power of Systems: How Policy Sciences Opened Up the Cold War World* (Ithaca, NY: Cornell University Press, 2016).

131. See the discussion in Lyubov G. Gurjeva, "Early Soviet Scientometrics and Scientometricians" (master's thesis, Universiteit van Amsterdam, 1992).

132. Like the reports Dobrov was analyzing, most of the forecasts produced by the center were not widely circulated. Nevertheless, he gained international recognition, mainly through his role as a savvy internationalist. In the 1970s, as the Soviet representative to UNESCO, he traveled widely in Europe and Asia, serving as a consultant for newly independent governments in the Third World. For instance, he consulted with the Iraqi government on setting up the Iraqi National Science Foundation. See G. M. Dobrov, *Irak: Tseli i sredstva gosudarstvennoy politiki v otnoshenii nauki*, Preprint-72-80 (Kiev: AN SSSR Institut Kibernetiki, 1972). From 1976 to 1978, he worked at a host of international institutions, including the International Institute of Applied Systems Analysis (IIASA) in Austria. On the IIASA and Soviet scientific internationalism, see Rindzevičiūte, *The Power of Systems*. On Dobrov, see Gurjeva, "Early Soviet Scientometrics and Scientometricians."

133. Dobrov, in turn, regularly visited the ISI in Philadelphia. Gurjeva, "Early Soviet Scientometrics and Scientometricians."

134. At the same time, computers were embraced by many groups within the counterculture movement. See, e.g., David Kaiser and W. Patrick McCray, eds., *Groovy Science: Knowledge, Innovation, and American Counterculture* (Chicago: University of Chicago Press, 2016).

135. Kelly Moore, *Disrupting Science: Social Movements, American Scientists, and the Politics of the Military, 1945–1975* (Princeton, NJ: Princeton University Press, 2008). For a more general discussion, see Jon Agar, "What Happened in the Sixties?" *British Journal for the History of Science* 41, no. 4 (2008): 567–600.

136. In 1968, the computer center at the University of California, Santa Barbara, was seized during student protests against the systemic racism of American universities. The result was the establishment of the university's Black Studies Department. See https://www.blackstudies.ucsb.edu/about.

137. See Paul Erickson, Judy L. Klein, Lorraine Daston, Rebecca Lemov, Thomas Sturm, and Michael D. Gordin, *How Reason Almost Lost Its Mind: The Strange Career of Cold War Rationality* (Chicago: University of Chicago Press, 2013).

138. See Benjamin Peters, *How Not to Network a Nation: The Uneasy History of the Soviet Internet* (Cambridge, MA: MIT Press, 2016).

139. See John F. Harris and Stephen K. Stearns, *Understanding Maya Inscriptions: A Hieroglyph Handbook* (Philadelphia: University of Pennsylvania Museum of Archaeology and Anthropology, 1997). On the history of Knorozov's deciphering of the Mayan hieroglyphic language, see Galina G. Ershova, "Teoreticheskoe nasledie Yu. V. Knorozova: K 70-letiiu pervoi nauchnoi publikatsii," *Izvestiia RAN* 77, no. 6 (2018): 60–68.

140. "The Problem of Deciphering Ancient Maya Writing," *Vestnik istorii mirovoi kul'tury* 1 (1957): 211.

141. I. M. Garskova, "Kvantitativnaia istoriia 1960-kh–1980-kh gg. v SSSR i ee rol' v stanovlenii istoricheskoi informatiki," *Istoricheskaia informatika* 3 (2018): 7–24.

142. For a contemporary review of the literature on the subject, see B. N. Mironov and Z. V. Stepanov, *Istorik i matematika* (Moscow: Nauka, 1976). For an overview of the trend, see 'Garskova, "Kvantitativnaia istoriia 1960-kh–1980 -kh gg. v SSSR i ee rol' v stanovlenii istoricheskoi informatiki."

143. See Mario Biagioli and Vincent Antonin Lépinay, eds., *From Russia with Code: Programming Migrations in Post-Soviet Times* (Durham, NC: Duke University Press, 2019).

EPILOGUE

1. Francis Fukuyama, "The End of History?" *National Interest* 16 (1989): 1–18, 7, 10, 17. For an elaborated thesis, see Francis Fukuyama, *The End of History and the Last Man* (New York: Free Press, 1992).

2. David Christian, "The Longest Durée: A History of the Last 15 Billion Years," *Australian Historical Association Bulletin* 59–60 (1989): 27–36. For a discussion of Big History as a genre of popular science literature, see Ian Hesketh, "The Story of Big History," *History of the Present* 4, no. 2 (2014): 171–202. For the history of scientific epics, see Zakariya, *A Final Story*.

3. David Christian, "The Case for 'Big History,'" *Journal of World History* 2, no. 2 (1991): 223–38, 227. See also Christian, *Maps of Time*.

4. D. Kristian, "K obosnovaniu 'Bol'shoi (Universal'noi) istorii," *Obshchestvennye nauki i sovremennost'* 2 (2001): 137–46. In the late 1990s, *Obshchestvennye nauki i sovremennost'* featured Russian translations of the work of another early advocate of Big History, Fred Spier. See F. Spir, "Struktura Bol'shoi istorii: Ot Bol'shogo vzryva do sovremennosti," *Obshchestvennye nauki i sovremennost'* 5 (1999): 152–63. David Christian's latest book, *Origin Story: A Big History of Everything* (New York: Little, Brown, 2018), was immediately

translated into Russian. See David Kristian, *Bol'shaia istoriia: S chego vse nachinalos' i chto budet dal'she* (Saint Petersburg: Azbuka-Attikus, 2018).

5. "Big History Center in Russia," *International Big History Association Newsletter* 1, no. 3 (2011): 1–6. The center was staffed by advocates of quantitative historical methods as a tool for political forecasting. Two members of the Euro-Asian Center for Megahistory and Systems Forecasting, Leonid Grinin and Andrey Korotayev, are affiliated with the Laboratory for Monitoring the Risks of Social-Political Destabilization (Nauchno-uchebnaia laboratoriia monitoring riskov sotsial'no-politicheskoi destabilizatsii) at the Higher Institute of Economics in Moscow (https://social.hse.ru/mr).

6. For a representative English-language publication produced by the center, see Leonid E. Grinin, Andrey V. Korotayev, and Barry H. Rodrigue, eds., *Evolution: A Big History Perspective* (Volgograd: Uchitel', 2011).

7. See https://www.sociostudies.org/news/1962673.

8. Christian has never publicly suggested a connection between his background as a historian of Russia during the Cold War and his turn to Big History at the moment the Cold War seemed to be ending. In an interview, however, he noted: "I taught [Russian history] during the Cold War when it seemed exceptionally significant. Teaching it in Australia, where I was, was a bit like talking about the dark side. . . . Having lived through the Cuban Missile Crisis, I remember it vividly. I was a schoolboy in England where this tribalism threatened to blow us all up. That was a very vivid experience for me." The dissolution of the Soviet Union might have left him searching for a new topic. See "We Need a Modern Origin Story: A Big History: A Conversation with David Christian," *Edge*, May 21, 2015, https://www.edge.org/conversation/david_christian-we-need-a-modern-origin-story-a-big-history.

9. See Serguei Oushakine, *The Patriotism of Despair: Nation, War, and Loss in Russia* (Ithaca, NY: Cornell University Press, 2009).

10. In the early 1990s, journalists, citizens' groups, and societies focused on recording and preserving the memory of the Soviet political persecutions (e.g., Memorial; see Nanci Adler, *Victims of Soviet Terror: The Story of the Memorial Movement* [Westport, CT: Praeger, 1993]).These groups and, to a lesser extent, foreign historians largely replaced Soviet-trained historians as public commentators on the meanings of the country's past. For the rapid changes in the ethics of journalism from truth seeking to the ethos of "post-truth," see Natalia Roudakova, *Losing Pravda: Ethics and the Press in Post-Truth Russia* (Cambridge: Cambridge University Press, 2017).

11. After the dissolution of the Soviet Union, alternative histories proliferated. See Kare Johan Mjør, "Smuta: Cyclical Visions of History in Contemporary Russian Thought and the Question of Hegemony," *Studies in East European Thought* 70 (2018): 19–40; and Marlène Laruelle, "Conspiracy and Alternate History in Russia: A Nationalist Equation for Success?" *Russian Review* 71, no. 4 (2012): 565–80.

12. A representative publication in English is A. T. Fomenko and G. V. Nosovskiy, *History: Fiction or Science?*, trans. Mikhail Yagupov, 4 vols. (Paris: Mithec-Delamere, 2003–6). The literature on Fomenko in Russian is vast. See, e.g., I. A. Nastenko, *Antifomenkovskaia mozaika* (Moscow: SPSL, 2001); and

S. O. Shmidt, *Fenomen "Fomenko" v kontekste izucheniia sovremennogo obshchestvennogo istoricheskogo soznaniia* (Moscow: Nauka, 2005). For a discussion of Fomenko and Fomenkoism in the larger context of Soviet and post-Soviet cultural politics, see Marlène Laruelle, *Russian Nationalism: Imaginaries, Doctrines and Political Battlefields* (London: Routledge, 2018); and Joseph Kellner, "The End of History: Radical Responses to the Soviet Collapse" (PhD diss., University of California, Berkeley, 2018).

13. See, e.g., A. T. Fomenko, "Nekotorye statisticheskie zakonomernosti raspredeleniia plotnosti informatsii v tekstakh so shkaloi," *Semiotika i informatika* (VINITI) 15 (1980): 99–124. On VINITI, see chapter 6 above.

14. Fomenko and Nosovskiy, *History: Fiction or Science?*, 1:i.

15. See Kellner, "The End of History." For a discussion of the erasure or downplaying of the Soviet era in Russian history in post-Soviet textbooks, see Vera Kaplan, "The Vicissitudes of Socialism in Russian History Textbooks," *History and Memory* 21, no. 2 (2009): 83–109.

16. Lawrence Bush, *Knowledge for Sale: The Neoliberal Takeover of Higher Education* (Cambridge, MA: MIT Press, 2017).

17. See, e.g., Jenni Conrad, "The Big History Project and Colonizing Knowledges in World History Curriculum," *Journal of Curriculum Studies* 51, no. 1 (2019): 1–20; and Andrew Ross Sorkin, "So Bill Gates Has This Idea for a History Class . . . ," *New York Times*, September 5, 2014.

18. See, e.g., Heikki Patomäki and Manfred Steger, "Social Imaginaries and Big History: Towards a New Planetary Consciousness?" *Futures* 42 (2010): 1056–63; Katherine Edwards, "Why the Big History Project Funded by Bill Gates Is Alarming," *The Guardian*, September 10, 2014; Sorkin, "So Bill Gates Has This Idea"; and Conrad, "The Big History Project."

19. National Science Foundation Research Reports, pt. 1, Kuhn Papers, box 20, folder 12.

20. See, e.g., Trevor Pinch, "Kuhn—the Conservative and Radical Interpretations: Are Some Mertonians 'Kuhnians' and Some Kuhnians 'Mertonians'?" *Social Studies of Science* 27, no. 3 (1997): 465–82. On the recent revisitation of *The Structure of Scientific Revolutions* and its historical contexts, see Robert J. Richards and Lorraine Daston, eds., *Kuhn's Structure of Scientific Revolutions at Fifty: Reflections on a Science Classic* (Chicago: University of Chicago Press, 2016). As Lorraine Daston and Sharon Marcus aptly put it, *The Structure of Scientific Revolutions* is representative of what they call *undead texts*—the texts that "remain crucial to the history of their fields" but whose "arguments and methods are no longer seen as viable." See Lorraine Daston and Sharon Marcus, "The Books That Wouldn't Die," *Chronicle of Higher Education*, March 17, 2019, https://www.chronicle.com/article/The-Books-That-Wouldn-t-Die-/245879.

21. See, e.g., John H. Zammito, *A Nice Derangement of Epistemes: Post-Positivism in the Study of Science from Quine to Latour* (Chicago: University of Chicago Press, 2004).

22. National Science Foundation Research Reports, pt. 1, Kuhn Papers, box 20, folder 12.

23. Kuhn, *The Structure of Scientific Revolutions*, 111. George Reisch has argued that the Cold War political context uniquely shaped Kuhn's thinking. See

George A. Reisch, *The Politics of Paradigms: Thomas S. Kuhn, James Bryant Conant, and the Cold War "Struggle for Men's Minds"* (Albany: State University of New York Press, 2019).

24. National Science Foundation Research Reports, pt. 1, Kuhn Papers, box 20, folder 12.

25. "Review of T. Kuhn's 'Philosophy of Scientific Development,'" National Science Foundation Research Reports, pt. 2, Kuhn Papers, box 20, folder 13.

26. "Panel Summary," National Science Foundation Research Reports, pt. 2, Kuhn Papers, box 20, folder 13.

27. See, e.g., George Steinmetz, ed., *The Politics of Method in the Human Sciences: Positivism and Its Epistemological Others* (Durham, NC: Duke University Press, 2005). For an account unpacking and historicizing labels such as *positivism* and those involving other *isms* often used in the historiography unproblematically, see Joel Isaac, *Working Knowledge: Making the Human Sciences from Parsons to Kuhn* (Cambridge, MA: Harvard University Press, 2012).

28. Quoted in Novick, *That Noble Dream*, 526.

29. Müller-Wille, "Claude Lévi-Strauss on Race, History, and Genetics."

Index

Note: Page numbers in italic indicate illustrations.